Cracking the Gender Code:

Who Rules the Wired World?

Women's Issues Publishing Program
Second Story Press

CRACKING THE GENDER CODE:

WHO RULES THE WIRED WORLD?

by

Melanie Stewart Millar

WOMEN'S ISSUES PUBLISHING PROGRAM
SECOND STORY PRESS

SERIES EDITOR: BETH MCAULEY

CANADIAN CATALOGUING IN PUBLICATION DATA

Stewart Millar, Melanie, 1970–

Cracking the gender code : who rules the wired world?

Includes bibliographical references and index.
ISBN 1-896764-14-2

1. Women - Effect of technological innovations on.
2. Technology - Social aspects. 3. Technological innovations -
Social aspects. 4. Women in technology I. Title.

HQ1233.S73 1998 303.48'3'082 C98-932047-2

Edited by Beth McAuley
Copyedited by Deborah Viets

Second Story Press gratefully acknowledges the assistance of the Ontario Arts Council and the Canada Council for the Arts for our publishing program. We acknowledge the financial support of the Government of Canada through the Book Publishing Industry Development Program for our publishing activities.

Printed and bound in Canada on acid-free, recycled paper

Published by
SECOND STORY PRESS
720 Bathurst Street, Suite 301
Toronto, Ontario
M5S 2R4

For Darius

Contents

Acknowledgements

THIS BOOK could not have been written without the support, criticism and skilled editing of Beth McAuley. I would go on at length, but I fear she would caution my over-sentimentality (!). Deborah Viets provided a very helpful and detailed copyedit of the final version. In addition, I have been blessed with a wonderful working relationship with the talented staff of Second Story Press.

Cracking the Gender Code had its beginnings in research conducted in the Graduate Programme in Women's Studies at York University and was completed in the midst of my doctoral study in Political Science. As a result I owe a great deal to the many wonderful professors and students at York who have guided and supported my work in crucial ways. I would also like to express my appreciation for the doctoral financial support provided by the Social Sciences and Humanities Research Council and thank the students in my tutorials for their frankness and for reminding me, on a weekly basis, why I was writing this book.

On a personal level, I am forever indebted to those who have inspired and encouraged me by sharing their strength, wisdom and laughter. These include Lisa Dale, Marlene Rumenovich, Larissa Silver, Jane Will, Diane Stadnicki, Varpu Lindström, Liz Millward, LaDonna and Allan MacRae, Scott Milne, Kylie Anderson, Deidre Martin, Lisa Littlefield, Laura Buckland, George Seremba, Kenton Bird and Gladys Johnston. Special thanks are due to Elisabeth Langford, whose encouraging cards, e-mails, phone calls and visits always came at exactly the right time. My second family, the Millars, have also been endlessly patient and supportive as a I continue to "do the academic thing."

It is impossible to adequately acknowledge my family's encouragement and faith in me as I was growing up and while I was writing this book. I continue to learn a lot from my parents, Sandra and Donald Stewart, as I flail about in this world. However, I will always be especially grateful to my mother for showing me feminism-in-action and for her gentle reminders about "balance," and to my father for his caring politics and shining example of integrity. I must also thank all my "sisters" — Jillian Stewart, Carrie Stewart and Jennifer Cameron — for their many insights and valued solidarity as we, each in our own way, "take on the world." I am your biggest fan.

This book is dedicated to my loving and supportive partner in all things, Darius Millar. For all you do and all you are, words are simply insufficient.

Note to Readers

Despite repeated attempts, I was unable to secure permission to reproduce many of the images and advertisements discussed in this book. While some corporate copyright holders denied my requests outright, others would grant permission only on the condition that they be allowed to approve the accompanying text or be assured that their image would not be portrayed "in any negative context whatsoever." These conditions restrict academic freedom and place unacceptable limitations on the constructive criticism of corporate advertising and images in popular culture. As a result, for many of the graphics analysed in this book, I have only been able to provide a description of the image in question and information about where it can be located.

Cover images and the text for some articles can be found by selecting the "Search Archive" option on *Wired*'s Web site at <http://www.wired.com>. Illustration credits can be found in full on page 230.

— *Melanie Stewart Millar*

Introduction

CULTURE CODES

> The impossibility of getting a grip, and grasping the changes underway is itself one of the most disturbing effects to emerge from the current mood of cultural change ... the revolutions in telecommunications, media, intelligence gathering, and information processing they unleashed have coincided with an unprecedented sense of disorder and unease ...
>
> — Sadie Plant, *Zeros + Ones: Digital Women + the New Technoculture*

AS I TYPE the characters forming these words, my fingers interrupt the impatient, flashing cursor that is forever ready for my next command. If I choose, the machine behind the screen will alert me to any spelling errors, typos and grammatical mistakes I make along the way. Should I require assistance, I can use the services of a help function that is never too busy for me, or access friends and colleagues through electronic mail. A world of information is accessible through my modem should I wish to use online resources and databases. I am connected to a futuristic "wired world" — a world that seems brimming with possibilities. Alone, with my computer at the ready, at any time of day or night, I can explore, research, write, communicate or entertain myself, in comfort and security. My options seem virtually endless.

In an apartment not far away from mine, a woman is answering the phone, taking an order for pizza and efficiently keying the information — name, address, size, toppings, deep dish or crispy crust? — into the pizza franchise's computer network. She glances repeatedly toward the kitchen stove to ensure that her own dinner has not boiled over. Her attention is divided by the multiple responsibilities that consume her life, all of which weigh heavily on her mind, like the growing stack of bills waiting to be paid. One of her children watches television while another, seated in a nearby playpen, mimics her mother's actions with a brightly coloured Fisher-Price phone.

Much farther away, in what has been described as another world, a fifteen-year-old girl concentrates intently on her delicate work. Her small fingers skillfully assemble the circuitry that will soon inhabit the next generation of personal computers. She works quickly for little money and for long hours. She focuses on what this money will mean to the brothers and sisters who remain in the village she has left behind. Hers is a race against time, as her eyes grow weaker and her fingers become less nimble under the strain of the repetitive action and close work.

These women, too, are the "beneficiaries" of the new wired world. Believed to be fortunate to have any work at all, the dawn of the digital age has meant very different realities for each of them, and for me. At my computer now, with the world apparently only a click of my mouse away, the seductive messages of television advertisements heralding the beginning of an exciting new interactive world seem almost plausible. So do the Toronto subway billboards that remind me that "there's lots of stuff in computers like microchips, wires, and oh yeah, a future."[1] It is easy to be swept away by this wave of unprecedented change, of dizzying, ever-accelerating technological transformation.

In fact, regardless of the media to which we look today, there can be little escape from the language of technological revolution and graphic depictions of a much-anticipated (if at times worrying) future. It has become a cliché of our time to speak of the shift from an economy based on industry to one based on information, and to muse about the marvels that such a change continues to bring. Since the early 1990s, the acceleration of information technology development and the growth of the Internet have intensified public perceptions of rapid technological progress. Cinematic and televised science fiction have melded with the "real life," "real time" marvels of an Internet computer age, to create a whirlwind digital spectacle of seemingly endless potential and awesome speed. Visions of a sort of futuristic, digital Disneyland abound as western culture rides the latest wave of technological wonder and euphoria. Vast public and private resources have been invested to institutionally package, advertise and ultimately sell these new technologies and the bright new age they have come to represent. As a result, the imagination of popular culture has been captured by evocative images of technological freedom and interactivity in an increasingly wired world. Metaphors such as the "information highway," "surfing the Net," "cyberspace" and "virtual reality" have moved out of the realm of science fiction, into the everyday world of the computer industry elite and computer hackers, and into the collective consciousness of diverse business people, educators, policymakers and the mainstream public.

Promises of technological progress not only stimulate technotopic fantasies and generate consumer excitement, but are also continually held out to us as the ultimate panacea for all manner of social ills. Such "technological bullets" have become significant tools of economic, cultural and political persuasion, as leading corporate giants promise to "deliver it all" in slogans that are now ubiquitous on our cultural landscape. IBM purports to provide "Solutions for a small planet," AT&T insists that "It's all within your reach," NEC Technologies challenges us to "Expect more. Experience more," and Microsoft repeats a single, haunting question of limitless potential: "Where do you want to go today?" Messages like these are underwritten by an ever-present, almost desperate, sense of urgency as the battle to survive in a perpetually "new" information economy reaches a fever pitch. The urgent demand for all of us to get "plugged in," "logged on" and "wired" is palpable, omnipresent, relentless. It would seem that western postindustrial society has become addicted to a sort of technological Prozac; a happy pill that promises to treat (or at least mask) the symptoms of capitalist society's chronic malaise — unemployment, social decline, urban decay, cultural alienation, vast economic and social inequality. The future, we are simply and repeatedly told, is digital — a veritable "technotopia" in which only the fully wired will survive.[2]

Meanwhile, politicians from Bob Dole to Newt Gingrich to Bill Clinton to Jean Chrétien trumpet the importance of the information technology sector in the global, competitive economy of the twenty-first century. Newspaper and television features echo this theme as advertisements for technical institutes remind us that we alone are responsible for our future employability in this new high-tech world. If we are skeptical about the ability of corporate capital to deliver the wonders of a "digital Nirvana," we must certainly believe that new information and communications technologies are our only hope for survival in an economic climate of international free trade and globalization. If our sense of adventure and consumer love of novelty don't motivate our acceptance of rapid technological change, our fear of being left behind will ensure that we eventually take the plunge.

However, before diving into the waves of a technotopia that seeks to engulf us, it might not be such a bad idea to test the waters first. In the context of a society that continues to perpetuate sexist traditions, ideologies and institutional practices, women in particular must gauge the incoming tide carefully. We must examine the waves that crash against the shore and inspect the foam that remains on the beach; we must scan the horizon for the source of each new wave, and understand the rhythms that regulate the tidal ebb and flow. We need to ask ourselves some very important and difficult questions about our relationship to the vision of technotopia that is presented to us, and how the realities of the current wave of technological change affect different women differently. We need to recognize the relationships between those who produce, consume, service and promote digital technologies. In other words, how does my experience of the power of digital technology relate to that of the pizza delivery phone representative or the computer-chip factory worker? Despite, or perhaps even because of, the dizzying pace of current technological change, we need to critically engage with our social context and determine how women's relationship to digital technology has taken shape, and how it may continue to develop in the future.

Women, Technological Exclusion and Strategy

In contemporary western culture, men are assumed to make the machines, and, if culturally appropriate, women may use them.[3] There is an unavoidable common sense to this observation; after all, what could be more closely identified with traditional images of masculinity than the technological "Progress of Man" from so-called barbarism to civilization? Such an identification is reinforced by millennia of historically constituted gender constructions that have come to define our very notions of what it is to be male and female. This relentlessly dualistic symbolic order has been repeatedly redeployed and vigorously — at times violently — defended in the history of western culture and political thought: the masculine is associated with reason, science, culture and production in the public sphere; the feminine, with passion, nature and reproduction in the private (or domestic) sphere. So deeply ingrained are such notions of male and female roles, that despite the many gains of the women's movement, western popular culture continues to find it difficult to separate biology from destiny: men are from Mars, women are from Venus.[4] Women remain largely responsible for minding the home and hearth (whether or not they engage in paid employment), while men are expected to conquer new worlds, make amazing scientific discoveries and transcend the mundane through high art and technological progress. Indeed, the goal of transcendence — the pursuit of human endeavour beyond the simple sustenance of life — has been historically articulated as an almost singularly masculine project. Women, meanwhile, are identified with what is viewed as the drudgery of immanence, forever preoccupied with the necessities of daily life (cooking, cleaning, reproduction and so on).

Why is this so? How did technology come to be thought of as the purview of men and connected to other traits of masculinity? The answer is not only controversial but also historically complex. In fact, telling the story of women's historical relationship to technology involves piecing together fragments of archaeological data, which have only recently been recognized as significant, and extrapolating from existing evidence. Piecing together the story demands that we abandon our assumptions that gender roles are historically stable and realize the extent to which our common understandings of human history are influenced by the sexist assumptions that exist today.

A pioneering book entitled *Mothers and Daughters of Invention: Notes for a Revised History of Technology*, by technology historian Autumn Stanley, provides one possible way to reconstruct this history in a way that accounts for women's social subordination and their exclusion from technology. Stanley traces human technological history from prehistoric times to the present and argues that, based on available archaeological evidence, it is most probable that the relations between the sexes were largely egalitarian in proto-human and early gathering-hunting societies. This was despite the fact that women's child-bearing role necessitated a sometimes more sedentary existence for them than for men. Such egalitarianism, however, was far from stable and was disrupted by the socially-constructed effects of successive technological developments. She convincingly argues that men and women devised different technological advancements based on the different roles they played in early tribes and societies. These roles, in turn, were then influenced by the way these inventions created social advantages for one sex or the other. A period of male advantage, she suggests, was gained as a result of the invention of projectile hunting (hunting with slings, stones, spears and eventually bows and arrows) in the Upper Paleolithic period. This was followed by an interval of female advantage in the early Neolithic period, when women invented the agricultural techniques that became central to the survival of early society. During this time, some cultures, like that of the Çatal Hüyük in what is now Turkey, reflected women's increasingly significant role through the worship of dominant female deities and the improvement of women's social status and living conditions. During the Bronze Age, however, many of these Near Eastern cultures that gave rise to our own experienced what Stanley refers to as "the Takeover." For a variety of possible reasons — including a population explosion, technological advances in agriculture and irrigation, the professionalization of agriculture, the development of long-distance trade and markets and the influence of nomads who worshipped male deities — women's status and social power declined sharply. As men began to dominate technology and technological invention, so they began to dominate the social order. This relationship between technology and social power has endured throughout much of the modern history of the West.[5]

It is important to recognize that women have not simply been the passive victims of technological change since this apparent "Takeover." Women have invented, adapted, theorized and even protested against technology throughout history. Women are now widely recognized as the inventors of agriculture and their inventions are known to have spanned the entire range of human endeavour, from medicinal and health-related technologies to reproductive technologies to industrial and domestic machines.[6] In fact, Stanley argues that of the five primary simple machines — the lever, the wedge, the screw, the pulley and the inclined plane — "women used and probably invented all or most of them in prehistoric or early historic times."[7] Yet women's systematic exclusion from the public sphere and from the institutions that support science and technology has ensured that technology remains a masculine domain. In this way, masculine social privilege and power have been reasserted with varying degrees of intensity throughout modern history. From the burning of female witches at the stake during the Middle Ages, to the exclusion of women from universities, to the rise of female invalidism and domesticity in the nineteenth century, women have been continually excluded from the construction of western scientific knowledge and technology.[8]

Nevertheless, a minority of exceptional women did contribute to the development of technology. They did so despite many obstacles, including legal impediments to patent protection and the ongoing lack of financial and social support, opportunity and access to technical and theoretical education. Unfortunately, the tenacity of these women has frequently been rewarded with obscurity. Their ideas have often been co-opted by male counterparts or ignored by historians who have either overlooked or willfully neglected female contributions to science and technology.[9] So, too, have many failed to recognize the way in which women's use of technologies, like the telephone, have influenced how technologies are adapted and developed in society.

Fortunately, the theoretical significance of technology to relations of gender in western social and political life has not escaped the notice of many feminist theorists. Even as long ago as the fifteenth century, for example, political theorist Christine de Pisan noted the importance of tools and architecture to gender relations in her *Book of the City of Ladies*.[10] Simone de Beauvoir, the celebrated "first feminist" of the Second Wave, also argued that as man "creates new instruments, he invents, he shapes the future" for both sexes.[11] While it has certainly been true that exclusion has characterized much of women's culturally defined relationship to technology, to suggest that women have been completely absent from technological history would be to obscure many important contributions and theoretical insights. Not only would such a view perpetuate the erasure of a history that women's studies scholars and historians have only recently begun to uncover, but it would also reinforce a particularly narrow notion of what constitutes technology. In other words, to claim that women's sole relationship to technology is one of exclusion would also be to imply that women's skills and knowledge in the private sphere (cooking, household work and so on) are not valuable or sophisticated enough to be called technology. Indeed, as Stanley's impressive documentation of female inventors indicates, if we expand our view of what constitutes technology to include inventions and processes often associated with the private sphere — such as weaving, pottery, cooking and childcare — women's contributions cannot be so easily overlooked.[12]

In light of women's historically problematic relationship to technology, it is perhaps not surprising that the history of women and digital technology is far more complex and intimate than is generally believed. The significant contributions of Ada Lovelace to the development of the Difference Engine and the Analytical Machine (early forerunners of the modern computer) support the fact that women did contribute to the development of digital technology.[13] Yet Lovelace's status as quite possibly the world's first computer programer is obscured by the legacy of Charles Babbage, who is credited with being the sole inventor of both proto-computers, and, as a result, the "father" of the modern computer. Grace Hopper also played an important, though often neglected, role in recent computer history. She invented the first computer language composed of words (FLOW-MATIC) and automatic programing, created and developed operating systems for early Mark I and Mark II digital computers and invented virtual storage.[14] Even the term "computer" itself was coined to describe the function of women who performed calculations and who wired hardware for the first digital electronic computer, the Electronic Numerical Integrator and Computer, or ENIAC.[15] While these contributions may be described as more exceptional than typical, it has even been argued by some that women have played the central historical role as "the simulators, assemblers, and programmers of the digital machines."[16] This claim may imply that women have a greater degree of power and agency in the modern computer industry than evidence can support. Nevertheless, one cannot dispute the fact that some women have played important roles in all aspects of the computer industry, including the design of hardware, software and artificial intelligence, as well as in the application of various programs.[17] Many more women have taken important steps to engage with current developments in digital technology in positive ways.

In fact, far from playing digital damsels in distress, many women are embracing the challenges of the Internet. These women are acquiring the skills of new information technologies; they are going to classes to learn how to use the Net, they are logging on, e-mailing, exploring online resources and subscribing to newsgroups, chatrooms, list-servs, MUDs and MOOs.[18] Women's groups, libraries and research centres are providing online Web sites for feminist research, which have the potential to greatly increase access to such resources. In fact, information about a myriad of issues of particular interest to women can be found online — from traditional subjects such as cooking and fashion to feminist news and activist events. In addition, the Internet itself facilitates important links of communication for many women's groups and feminist organizations. Online groups like SYSTERS and Virtual Sisterhood, electronic "zines" like *geekgirl* and resources like Web-Networks' WomensWeb are encouraging women to overcome inhibitions about technology and recognize, in the words of Internet advocate and cyberfeminist St. Jude, that "Girls *need* modems."[19] This evidence suggests that the Internet is at least as responsive to women's issues (if not more so) than many other forms of media. It also has the potential to provide women with a unique forum for expression. Still, "woman-friendly" Internet resources remain in the minority as do the number of women online. It is extremely difficult to determine precisely how many women use the Internet. Estimates in the early and mid 1990s have varied widely from less than 1 percent[20] to 15 percent of all users.[21] This percentage does seem to be rising. One optimistic 1997 Nielsen Survey even claims that 42 percent of online users are women.[22] Yet, whatever the exact figure, it is clear that women — especially poor women, disabled women and women of colour — remain underrepresented online.

The apparent rise of women's online activity and the presence of feminist electronic resources have complemented (and in some cases inspired) new directions in feminist approaches to technology in the West. Although many of these approaches share much with traditional liberal feminism, others represent a more significant departure from dominant second-wave feminist understandings of technology. While earlier feminist theorists have been accused of leaving the impression that technology is "inherently patriarchal and malignant,"[23] the 1990s seems to have brought with it a more technophilic air. By providing personal anecdotes about how technically literate women can find alternative communities and entertainment online, many feminists are clearly recognizing the utility of digital technologies.[24] Articles of a similar ilk dominated the submissions I received for the "Women and Technology" issue of a feminist quarterly that I recently edited.[25] It is as though there is an almost cathartic value to the sharing of such personal accounts — a collective sense of reassurance in the knowledge that women can and are meeting the incessant social demand to get online and to get wired. And quickly.

Figure 1: Selling the laptop computer as the answer to the plight of the modern working mother. Who really benefits?

However, in this frenzy an alarming convergence is occurring — women's rush to get online so they're not left behind in the increasingly digital world of the twenty-first century is finding a disturbing resonance in the rhetoric of mainstream political and cultural leaders and in the advertisements of large computer corporations. This is starkly portrayed in a recent ad from Compaq computers (see Figure 1), which uses the idea that digital technology will allow western women to bridge caregiving and career responsibilities by facilitating home telework to sell its laptop computers. The ad simultaneously urges women to join the new information economy by consuming digital technology products and reminds them that their primary responsibilities to home and family will remain constant. They are told: "You can make sacrifices for your career. You can make sacrifices for your family. Or you can choose not to make sacrifices." The message here to male consumers is both familiar and deceptively simple: far from threatening the existing unequal gender relations, women's use of digital technology will actually preserve and perpetuate them. In this view, the demand for women to get online takes on a very different meaning than that assumed by feminists who urge women to move beyond technophobia and embrace the new opportunities provided by digital technology. The language of technological change found in the Compaq ad reinforces the concept that women are passive users of technology while at the same time reinforcing their traditional sex roles. The technologies of the future are allied with the traditional views of the past.

The difference between women encouraging other women to become users of digital technology and ads like Compaq's is clearly the context in which it is delivered. One is supported by the contextual history of the women's movement and the ongoing countercultural effort to improve women's status in a society that continues to socially, politically and economically discriminate against women. The other reinforces some of the very gender stereotypes against which women continue to struggle. And yet both implore women to get online. How can the resonance between such diverse voices, with such disparate levels of cultural power, be understood? What does this suggest about women's relationship to digital technology in contemporary culture and the possibilities for feminist intervention? In view of the traditional messages delivered by companies like Compaq, is computer savvy really enough to ensure women's success in the new digital age?

To answer each of these questions, we need to understand several things: the power relations in each context and the different scale on which such messages are disseminated; how power is circulated in our society through language, discourse, ideology and technology; and how the dominant (or hegemonic) messages of popular culture influence, and are influenced by, capitalist economic relations and enduring systems of sexism and racism. We also need to understand how new ways of perceiving and knowing the world are devised and generated throughout our society and how they become significant forces of persuasion. Only then can we move beyond the "unprecedented sense of disorder and unease" and begin to get a grip on the meaning of the rapid cultural change associated with the rise of digital technology.

Power, Discourse, Ideology and the Rise of Digital Culture

Everyday, we witness and experience the multiple power relations that circulate throughout society. When we say that an individual or a group is powerful or is exerting its power, we are recognizing the existence of a *power relation*: one entity exerting its ability to compel another to do, think or feel something it would otherwise not do, think or feel. All aspects of society are implicated in these power relations, which, though multiple and shifting, tend to follow particular, recognizable patterns. Such patterns become known as the dominant, or *hegemonic*, power relations of a given culture or society. In western culture, we can easily identify many of these power relations as reinforcing relations of domination and subordination: masculine over feminine, white over black, rich over poor, able-bodied over disabled and so on. Such hegemonic power relations provide an important template that can be used to help decipher how a given culture is to be interpreted. This is not to say that dominant power relations are capable of explaining why every single event occurs in a given society or that we are powerless to influence how particular incidents play themselves out. However, such power relations do manifest themselves frequently enough that they are easily identifiable as the norm rather than the exception. They are also the single most significant indicator of how society regulates complex human behaviours.

Hegemonic relations of power are reproduced in two main ways — *materially* (through physical threat, economic necessity or force) and *discursively* (through language, myths and ideological persuasion). Of course, neither of these methods exists independently of the other — discourse is reinforced by material conditions and vice versa. In our contemporary, media-saturated society, the influence of discourse over how people live, behave and think about their lives cannot be underestimated. While our socioeconomic, cultural and sexual status are key variables in how we interpret and are interpreted by the society in which we live, these categories are re-created and given meaning by the construction and circulation of discourse. Just as the collection of words, symbols and linguistic structures that the medical community uses shapes the way in which doctors, nurses and technicians understand their profession and location within the world, so do multiple social *discourses* influence how people come to experience themselves, other people and the world within a social context. We come to know ourselves and our place in the world through the narratives or discourses that flow in our culture: through television, film, advertisements, social interaction, the spoken and written word — even the structure of language itself — we learn to make sense of our lives. In this process, we may also develop, or come to support, specific ideologies, as a way of creating clarity and coherence. Jay L. Lemke describes ideology as "the very common meanings we have learned to make, and take for granted as common sense, but which support the power of one social group to dominate another."[26] Many different ideologies of varying degrees of popularity and scope exist in society and are articulated through particular discourses. Thus relations of power are re-created and perpetuated through discourses that reflect particular ideological beliefs.

In times of rapid technological change, like our own, the creation and circulation of new discourses and the ideologies that underlie them is accelerated. Widespread social and cultural changes are demanded by rapid and expansive technological change; new discourses help us to adjust to technological change and understand its significance. The development of digital technology and its rapid dispersion in Western industrial society is facilitated through the rise of an accompanying *digital discourse.* This emerging digital discourse, brought to us by leading technologists, computer industry elites and journalists, performs a number of important functions in society: it stimulates the social need for digital services, circulates myths of technotopia, popularizes new language and metaphors, and, of course, it sells digital hardware and software itself. As the Compaq advertisement for laptop computers suggests, digital discourse is also implicated in the social, political and economic climate of the larger culture. This discourse seizes upon and reflects particular relations of power through its accompanying ideology, and, as we shall see, reintroduces beliefs that perpetuate inequalities of gender, race and class. It may even explicitly or implicitly promote very specific social and political policies and discourage others. Digital technology itself, then, is enmeshed in a larger contextual framework, complete with its own emerging discourse and ideology. New information and communications technologies are never simply masses of circuitry, streams of ones and zeroes, or the intricate webs of the Internet. Rather, these digital technologies are part of a complex social and political context that profoundly affects our lives.

Cracking the Gender Code of Digital Culture

For these reasons, this book is not about the online culture of the Internet that we hear so much about, but about the net surrounding that online world. *Cracking the Gender Code* explores how digital technological change is being packaged and sold to the public through cultural messages that support a particular view of how the future should be organized. It looks at the false promises, delusions and violences of this emerging digital discourse that are remarkably well funded and promoted. It attempts to crack some gender code; to hack a way into this increasingly pervasive world of digital hype in order to reveal the underlying, profoundly sexist and racist systems of belief that it creates and perpetuates. Like genetic code, digital gender code facilitates its own reproduction, continually re-creating itself in a sea of complex and sometimes contradictory symbols and institutional practices. But it cannot do so without the right conditions. *Cracking the Gender Code* navigates a feminist path through these cultural symbols and conditions to discover what gets caught in the net of digital culture, what is simply ignored or manages to escape and who rules this new wired world. Most importantly, this book attempts to demystify the high-tech world that surrounds us so we can determine possible sites of future resistance and attempt to disturb those conditions that support the reproduction of digital gender code. *Cracking the Gender Code* is an effort to answer two deceptively simple questions: What is going on here, and what does it mean for women?

Chapter 1 begins this difficult task by mapping how existing social structures and cultural ideas support, and are supported by, the rise of digital technology, discourse and ideology. It examines the relationships between the emerging digital culture and the ongoing historical processes of modernity, the development of capitalism and shifting constructions of masculinity and femininity. The goal here is not to provide an explanation for every aspect of digital discourse but to provide a broad context that will help to demystify digital discourse and ideology. It will also act as an antidote to the hype and hysterical technophilia of the digital age. Chapter 2 explores how social commentators and western feminists are currently theorizing and understanding the rise of digital technologies in society. I explain how a feminist politics of anticipation can be used to demystify and resist the emerging discourse of digital culture and suggest that gender discourse analysis offers one important political tool that we can use to understand some of the directions that our digital society may be taking. The chapter also outlines the analytic tools to be used in this book and introduces *Wired* magazine as one of the most vivid examples of digital discourse. Chapter 3 explores *Wired* as a medium, situating it on the cutting edge of digital culture and ideology and describing how the *Wired* machine functions, and why it matters. Chapter 4 begins the work of analyzing *Wired* discourse, exploring how the magazine creates its exclusive, wired world — a world that excludes, reconstructs and ultimately tries to completely eliminate social difference. Chapter 5 explores the construction of the hypermacho man and the ideals of masculinity that are integral to much of the emerging digital ideology. Moving beyond these images of masculinity and otherness, Chapter 6 identifies and explores six of the most damaging myths of digital ideology as found in *Wired* and, increasingly, in other areas of popular culture. The relationship of these myths to existing power relations of inequality is examined. Finally, in Chapter 7, I discuss possible strategies that women can use to resist elements of digital ideology that are potentially oppressive and to reintroduce justice and responsibility into digital discourse. The extent to which digital technology can be viewed as both potentially liberatory and repressive is revisited in this final chapter, as we explore possibilities for feminist engagement with technological change in the next millennium.

Chapter One

TRACING THE ROOTS OF DIGITAL CULTURE

> So deep is society's collective ignorance of what cybertechnologies are doing to us that the first urgent task is to discover our ignorance ... And we can use our speculation of this weird future as a means of reflecting on ourselves and on the assumptions we have inherited about the way the world is, should be, and yet could be.
>
> — Ziauddin Sardar and Jerome R. Ravetz, *Cyberfutures*

> Over the last century the vocabulary has changed, perhaps, but the dominant mythic themes have not; they have remained intact, and all the more effective because unconscious.
>
> — David Noble, *Progress Without People*

> Ideas only become effective if they do, in the end, connect with a particular constellation of social forces.
>
> — Stuart Hall, "The Problem of Ideology: Marxism Without Guarantees," *Journal of Communication Inquiry* 10

DIGITAL DISCOURSE THRIVES on speed and novelty. With each new wonder of digital programing and each worker declared "redundant" by technological restructuring, we have come to see contemporary technological change as rapid, unforgiving and inevitable. Unfortunately, between the streams of digital data and the flashing cursor, there simply isn't time for critical thought. The cycle from the "next great thing" to obsolescence to the "next great thing" has accelerated to the point where the belief that we can keep up can only either reveal our ignorance or indicate a life lived in blissful denial. The message is clear: in this time of massive social change, the likes of which we are told western culture has never seen before, the most we can do about the future is wonder what lies ahead. The problem is that such passive speculation is valuable only as entertainment. It is as ephemeral as the images that flash across our computer screens, dazzle our television sets, occupy our dreams. Idly wondering about what the future will deliver to us next offers no critical edge and prevents us from actively engaging in the process of building the future together. As a result, mere speculation about the future of a technologically determined world threatens nothing and concedes everything: we can't understand the future or engage with it because *they* are still inventing it.

The sense of powerlessness that results from such discourse is not accidental. The technological and cultural changes that are now occurring have been discursively constructed in such a way as to appear completely unprecedented, dazzling and unstoppable. From within such a discourse of speed and novelty, critical understanding of what is going on *is* impossible. However, the discourse and ideology of digital technology are not beyond our comprehension. We simply need to step back and recognize digital discourse as a particular way of constructing and understanding our world. We need to give this discourse a history, recognize its connections to socioeconomic conditions and to other dominant ideas and understand how it influences our current understandings of power relations. In other words, we must perform those tasks that are well known to feminist theory: we must *remember* how our past has helped to create our present, *reconnect* the power relations that exist between culture, economics and technology, and *relearn* how to interpret and critically engage with the social discourses of our times. In doing so, we must liberate ourselves from our role as spectator and recognize that what lies at stake for all of us in the way digital discourse is constructed must not be so easily forsaken. It is, afterall, our future, too.

Remembering the Past: The Legacy of Western Science and Modernity

Human beings do not spring into life fully grown; neither do new discourses. Like humans, they develop over time and under the influence of larger social forces and modes of thought. These modes of thought influence decades, or even centuries, of human history. The development of digital discourse is influenced by multiple parental genes, scattered throughout western social and political thought and history. Digital discourse is neither entirely dependent on current realities nor completely novel, but bears a strong family resemblance to its ancestors. In addition to the legacy of women's relationship to technology before and after the "Takeover" (as discussed in the Introduction), the less distant relatives of digital culture must also be considered. One important strand of these parental genes lies in the history of the rise of western science and technology and the development of modern industrial society. Herein lie many of the origins of our contemporary technologically driven society. If we are to understand the current processes of technological change, economic restructuring and the emergence of digital discourse, it is useful to recall how these processes have functioned in the past.

The particular form of western science with which we are familiar has not always enjoyed the high status and virtually unquestioned authority that it is blessed with today. The emergence of science and technology as a dominant force has a long history. Enmeshed in this history is the social construction of these pursuits as the sole purview of men. According to historian David Noble, women's exclusion from the modern institutions of science in European societies can be traced back to the tenth century and the Christian clerical tradition. He argues that the origins of western scientific institutions and ideology lie in the Latin Church — an institution with a cultural tradition of celibacy and misogyny.[1] Noble contends that the "allegedly fundamental conflict between science and religion" is "greatly exaggerated" in common historical understandings of the scientific revolution of the seventeenth century.[2] Far from ushering in something entirely different from religion, technology came to be imbued with the same spiritual significance and masculine associations that religion once had. As a result, modern man began to seek the achievement of a sort of worldly spiritual salvation through technological developments rather than through a divine afterlife. In this way, science and technology replaced religion as the primary means through which modern western man sought transcendence.[3]

In order to establish the patriarchal authority over western science that is so crucial to the creation of this "masculine millennium,"[4] it was also necessary to eliminate traditional knowledges that may have served to undermine its superiority. The exclusion of women from the early institutions of science and technology alone was not sufficient to achieve masculine dominance in this transcendent field. Alternative knowledge systems had to be destroyed as well. As a result, the Catholic and Lutheran churches waged a war upon the community of wise women and women healers of Europe. Millions of women were burned at the stake as witches in a genocide that lasted for three hundred years, from 1400 to 1700. These women were sacrificed in order to protect the interests of the growing male medical establishment and to destroy all remnants of goddess-worshipping religions.[5] During this period, the dominant cultural image of nature shifted from one of a nurturing mother to one of an unruly, hysterical woman who required the mastery of man. Francis Bacon's utopian vision of an idealized world of science in which man dominates nature shows us the extent to which even the seventeenth century "father" of the scientific revolution promoted a misogynist, instrumental view of women and the natural world.[6] Both women and nature were seen to require the discipline and control of masculine science and technology.

The rise of western science and technology also played an important role in the construction of how racial difference is perceived in the West. After the scientific revolution, western belief in scientific and technological superiority over non-western cultures influenced how European ideologies of imperialism were shaped. "Machines" became "the measure of men" and were used as a way of distinguishing between and assessing the worth of different peoples. Scientific and technological standards of achievement were infused with a symbolic cultural significance during the colonial period. As a result, the religious mission to convert foreign peoples to Christianity was gradually replaced by a desire to "civilize" them through western science and technology.[7] Science and technology became the hallmark of western cultural superiority, important not only for the institutionalization of a male-dominated culture but also for the construction of western identity as superior to that of non-western peoples.

This dramatic elevation of the meaning and significance of what Noble calls "the useful arts" to the status of modern technology was central to the achievement of a modern social order. In fact, the project of modernity itself arose as an optimistic response to the growth of scientific knowledge and to the great social and economic changes that occurred during the seventeenth and eighteenth centuries, which culminated in the industrial revolutions of the late eighteenth and nineteenth centuries.[8] Modern thought sought to free humanity from scarcity by using science to regulate nature and to prevent political tyranny by using reason to engineer social organization. The darkness, mythology and mysticism of the Middle Ages would be superceded by a modern era of human reason in which science and technology would facilitate the progress of a universal human subject and the fulfillment of his potential.[9] Modernity firmly held to the truth of two main assumptions: (1) that the social (and physical) world is intelligible, and (2) that this world is malleable as a result.[10] Humanity's understanding of the natural and social world was expected to continue to grow exponentially, and, with such growth, the universal qualities of humanity and the natural world would be revealed. By patterning social organization after rational science, it was believed that "man" could embark on a historical path of progress that would enhance all areas of human life and reveal universal truths of justice and equality.

The modern capitalist society that emerged in the nineteenth century was partly a result of the industrial revolutions. Capitalism relied heavily on technological innovation to organize labour and extract profit from the new mode of production — namely, the industrial factory. This, too, was not achieved without violence. For example, the Luddites (who were weavers, named after their mythical leader, Ned Ludd) demonstrated their resistance to mechanization and fear of job loss by smashing mechanical looms. At the height of the Luddite rebellions in 1812, 24,000 British troops were needed to restore order.[11] The Luddite struggles were ultimately unsuccessful and, despite the onset of a harsh period of economic dislocation, industrial society continued to develop undeterred. With the growth of this industrial society came many of the characteristics that define modern life: a growing sense of individualism and alienation, a loss of community and a yearning for ideological substitutes, a sense of constant innovation and increasing urbanization. There also arose a particular economic and social structure.[12] With the division of the public sphere of the male citizen and male worker from the private, domestic sphere of the wife and mother came a specific, heterosexist model of gender relations. In the words of sociologist Barbara Marshall, "once industrialised, capitalist class society is introduced into history, women and children disappear."[13] This "disappearance" has resulted in the neglect of women's unpaid domestic and reproductive labour as significant elements in the survival of the capitalist economic system. It has also facilitated the continued identification of women as, at best, secondary or occasional workers, who, as a result, still only earn a fraction of male wages. The rise of the nuclear family and the disappearance of women in modern industrial society are of crucial importance to women's social and economic location today. The confinement of large numbers of women to private lives of economic subordination institutionalized modern gender relations and paralleled the exclusion of women from science and technology.

Many of these features of modernity have remained in place throughout the late twentieth century and, in fact, are very much present today. Science and technology continue to be dominant social, economic and cultural forces that promise to cure disease, provide answers to many of our most profound questions, expand the possibilities of human potential and increase production and profits. Technological progress is still a widely held and largely unexamined belief, and technological change has been generally accepted as an inevitable feature of our modern life. Technology itself is regarded as an essentially neutral tool, guided by a rational, universal science — both of which are independent from subjective values and economic systems. Technology continues to be defined as a masculine pursuit. The majority of our governmental and corporate institutions are still structured in a formal, highly rationalized bureaucratic order, and, although plagued by decline in an era characterized by the rising influence of multinational corporations, the modern state remains the site of ideals of public justice, democracy and equality. In addition, despite the presence of countercultural ecological movements, our practical response to the natural world continues to be one of disconnected exploitation. Meanwhile, nations of the South continue to be treated as "developing countries" in need of western technological knowledge and industrial and social organization. Digital discourse has grown up with and repeats many of these modern hegemonic assumptions. But it has also been influenced by the reactions to modernity that have arisen in the last three decades.

The End of Modernity?

The dominance of modernity in western societies was challenged in the 1960s by the rise of countercultural and antimodernist movements. These groups questioned the validity of the modern project itself and argued that modern society was oppressive. Criticizing the scientifically grounded, technical, bureaucratic rationality of state and corporate power, they demanded that modernity deliver the justice and equality that its belief in rational social organization promised. Critiques of modernity came from all sides, from feminists revealing the sexist nature of modern social organization, to antiracist movements recognizing the imperialistic, racist assumptions of modern western society. There seemed to be a "crisis" of modernity; a growing lack of faith in the rationality of a growing technologically driven society, and, in the wake of two horrific world wars and the emergence of the Cold War in the 1950s, the ability of the modern project to liberate humanity from tyranny. One result of this crisis was the emergence of a new theoretical paradigm, one that David Harvey dates between 1968 and 1972, known as postmodernism.[14] Yet despite the subsequent declaration by many theorists that a "postmodern condition" has existed in western industrial society since the late 1980s, the characteristics of such a condition remain in dispute. If we are, in fact, living in a postmodern era, what does that mean? And how has the emergence of postmodernity influenced digital discourse?

Perhaps the most important departure that postmodern thought makes from that of modernity is that it affirms what modernism tries to regulate. That is, rather than aspiring to provide an all-encompassing metanarrative in order to explain, rationalize and structure society, postmodern theory embraces fragmentation, chaos and change. Postmodernism argues that if we are to avoid the pitfalls of a repressive system that seeks to engage humanity in a global project patterned upon scientific rationality, we must become comfortable with more questions and fewer answers. Rather than articulating a "project," postmodernists describe society in terms of a "condition"; a way of being in which time is collapsed, multiple voices are embraced, difference is celebrated and identities are fragmented. The faith in abstract reason and the power of science and technology to emancipate the eternal human subject is lost in the postmodern world. So are all attempts to put forward a universal spiritual or moral goal for human life. The universal rational subject of modernity is destabilized and reinterpreted as fragmented, multiple and ever-shifting identities.

Postmodern thought has important, though very controversial, implications for western feminism. Feminist thought has relied on the existence of a unified subject and a common experience in order to articulate injustices against women and mobilize their resistance. By throwing the unified subject into dispute and suggesting that society is ultimately unknowable, postmodern thought seems to take the wind out of the possibility of collective political aspirations for equality.

However, postmodern theory has also provided feminism with some very important insights. It helps us to more fully understand the problems of using the category "woman" simplistically, without recognizing the multiplicity of experiences and subjectivities of different women. Of course, this criticism is not new and has been raised by women of colour who justifiably argue that western feminism has far too often been defined to meet the needs of upper-middle-class white feminists. As well, postmodern theory has provided feminism with some very powerful analytic tools. By emphasizing the role of language and discourse and by demonstrating the multiplicity of power relations that exist within society, postmodern thought has helped some feminists build more sophisticated theoretical understandings of society.

DIGITAL DISCOURSE AND HYPERMODERNITY

But can society's much discussed "postmodern turn" help us to understand the rise of digital culture? Some of the characteristics of digital discourse — discontinuity, speed, symbolic and linguistic spectacle, the tendency to exaggerate the novelty of the present — are clearly reminiscent of postmodernity. Yet digital discourse does not seem to represent a significant departure from the central elements of modernity. Far from the heady world of postmodernism, digital discourse seeks instability and indeterminacy only in so far as rapid technological change necessitates them. In fact, there remain continuous battles between existing elites to seize control of and stabilize the commercial potential of digital technologies as well as the potential of research and development. These battles have been so intense that the numerous business mergers and takeovers that have characterized the 1990s have been described as nothing short of a "corporate feeding frenzy over control of the information highway."[15] We may be experiencing a shake-up among existing corporate elites, but hegemonic power structures remain intact. And there can be little doubt that digital discourse maintains science and capital as organizing principles as it tries to sell the "digital life" to western consumer culture. Digital discourse does not simply turn its back on the master narratives of modernity. Nor does it destabilize the universal (male) human subject or distrust scientific reason. And, most emphatically, digital ideology does not valorize difference and heterogeneity. Whether we describe our world in terms of a modern or postmodern era, it is clear that patriarchal relations of domination continue undeterred.[16] It would seem that digital discourse intensifies those power relations with which we are already familiar, repackaging them in the postmodern aesthetic of the day.

This repackaged, accelerated form is intense and dazzling, and, perhaps more than any other discourse of our times, is infused with a hyperactive sensibility — a disjointed and frenetic style that incorporates some of the superficial aspects of postmodernity. As a result, it is useful to designate this emerging epoch as "hypermodern." This term reflects both the new characteristics of digital culture and the endurance (and intensification) of modernity's drive for scientific and technological progress in a capitalist context. As we explore this emerging digital culture, we must not become too distracted by the new features of hypermodernity that it possesses. The core of digital discourse remains thoroughly modern.

Reconnecting the Power Relations: The Needs of Global Info-Capital

The emergence of digital discourse and its composition cannot simply be explained in terms of shifting patterns of thought or fluid historical conditions of being. Discourse must be understood within the material conditions in which it is situated. In western industrial society and culture, this means coming to grips with the needs of capital and its role in the manipulation of social realities. Both postmodern and postindustrial theories often "make technological development the motor of social change and occlude ... economic imperatives."[17] Yet the emergence of new information and communications technologies and the rise of digital discourse cannot be separated from the logic of the mode of production that has facilitated their design, production and proliferation. In other words, digital discourse both supports and is supported by the needs of capital. The most important of these needs in recent times has been the transformation of the traditional, factory-oriented economy operating mainly within the nation–state, into an emerging digital economy that operates almost entirely within a global framework.

Capitalism, it is sometimes argued, has historically used technological change as a way of breathing new life into dying capitalist cycles.[18] In its latest incarnation in the form of multinational corporations, capital requires greater levels of profit extraction, enhanced marketing opportunities and increased mobility in order to thrive. New information technologies decrease the need for costly human labour by facilitating the replacement of many "information workers" with information machines. The results can be extremely lucrative for business. In Canada, for example, the Royal Bank has claimed that the record breaking profits that it has enjoyed in recent years have been due mainly to increased service charges derived from automated services like bank machines and telebanking.[19] Yet there can be little doubt that the accompanying reduction and reorganization of labour costs has also helped. It is important to recognize that despite claims to the contrary, the increased use of new information technologies in business has not resulted either in an improvement in the quality of the jobs available or an increase in the quantity of jobs. Rather, existing jobs have been destabilized, and frequently de-skilled, and overall unemployment rates have increased.[20] It is the lean, downsized corporations — made so through increased reliance on technology — that are the most profitable in the 1990s.

In an age of increasing globalization and multinational expansion, we also need to recognize the role of digital technology in the expansion of (primarily) American capitalism and the cultural, economic and political imperialism that accompany such expansion. It is clear that "while the 'Net is global in scope, it is undeniably run by an amorphous agglomeration of U.S.-based multinationals."[21] This process of global "modernization," facilitated by the use of military and communications technology, is similar to earlier forms of "civilizing" western imperialism. Indigenous cultures are subject to external influences that often destabilize, and adversely affect, existing ways of life. The result is the same as in previous waves of imperialism: "cultures are decimated, bulldozed, globalized with barbaric abandon."[22] But culture is not the only casualty when western multinationals "modernize" other countries and set up shop overseas. The findings of the Third World Network, an advocacy group for countries of the South, indicate that the microelectronics industry, on which all of digital ideology depends, has not eradicated unpleasant, labour-intensive work, but has led to a "degradation of skills, dignity and job interest for the majority of those working with the new technology" throughout the world.[23] Not surprisingly, much of the assembly of new information and communications technology is achieved through the exploitation of low-cost women's labour in the "developing world." In fact, the vast majority of workers in the world's Export Processing Zones (EPZs) — areas designed to attract foreign companies through tax-free incentives and suspended labour laws — are women.[24] Not only are these women subject to exploitative work conditions, but they are also routinely exposed to harmful toxins and fumes.[25] Although the impact of such exploitation has negative effects for many in the developing world, "it is women who bear the brunt of this suffering."[26]

Even these cheap female workers are becoming expendable to the computer industry. They too are easily replaced by a more automated manufacturing process. As more and more computer industry manufacturing is being mechanized, many of these low-paying, low-skilled jobs in the South are disappearing. The jobs that remain or are created as a result of such automation are found mainly in the male-dominated sector of technical, managerial and programing jobs — jobs that require education and training that is denied to the vast majority of women. The results are already showing up in a reduction in women's share of the employment in the Malaysian electronics industry.[27] It is not clear how the deplacement of these workers will be absorbed within these societies. Meanwhile, with the control of research and development into new information and communications technologies residing in the hands of a small number of powerful transnational corporations that remain focused on western consumers, it is unlikely that the specific needs of women of the South will be considered in the corporate strategies of the future.[28]

Digital technology has also enabled corporate capital to find new markets and new sources of profit by providing the means to reduce all knowledge, including culture, art and even DNA, to a stream of ones and zeroes — the binary language of computer networks. Digitization allows corporations to commodify more and more aspects of human life by reducing them to a common form and selling them electronically. For example, the latest developments in biotechnology have made possible the patenting and selling of genetic material. Multinational biotech and pharmaceutical companies are already engaged in a form of "biopiracy" as they mine and patent plants from the South for profit.[29] Meanwhile, the Human Genome Project, a program funded by the United States government to "map" human genes, may facilitate the commodification of human genes in the near future. With the help of a computerized database of biological information, parents could pre-select the genetic make-up of their children. In fact, according to Jeremy Rifkin's *The Biotech Century*, the primary purpose of new information technologies is to provide a means to co-ordinate and manage a new economy based on the manipulation and sale of genetic resources. This would help to explain why Microsoft, the U.S.-based software giant, is investing so heartily into this new field of bioinformatics.[30]

Another benefit to corporate capital is the great speed with which information and communications technologies can transfer monies and transact business. Funds can be transferred, production sites relocated and pertinent business information exchanged within minutes with the help of a seamless web of digital networks that are oblivious to national borders. The digital world is not only a smaller world, it is also one in which corporate opportunity is virtually unlimited and unrestricted. The new technologies have also spurred the rise of completely non-productive economic investment, particularly in the form of financial speculation. Fortunes are now made and lost without a single good or service being exchanged as the transfer of "digital money" is made between stock exchanges from New York to Hong Kong. Digital technology has been "instrumental" in facilitating the "restructuring of the capital system from the 1980s onward."[31] In the vernacular of the new high-tech discourse, it would seem that the future of business really is digital.

The many demands of info-capital have resulted in significant social changes in western industrial societies. Two of the most important of these have been the implementation of international free trade agreements and the devolution of the role of publicly accountable nation–states in managing domestic economies. The latter is evident in what Linda McQuaig has referred to as the "Cult of Impotence," which has provided the ideological basis for the divestiture of western nation–states from responsibility for the economy.[32] At the same time, states have played an increasingly important role in facilitating the mobility of capital internationally. This is highlighted by the negotiations that recently took place among members of the Organization for Economic Cooperation and Development (OECD) to secure a Multilateral Agreement on Investment (MAI). This proposed agreement, which has been referred to by critics as nothing short of a declaration of "global corporate rule," is designed to ensure the undeterred, international free flow of capital investment.[33] If successfully enacted, the treaty may prohibit any government restriction on such capital flows, including the regulation of corporate profits. Why is such a treaty being pursued now? After all, foreign investment has become a common feature of the increasingly globalized form of capitalism with which we are familiar. The rise of the MAI may also be linked to the development of new information and communications technologies. Not only do these technologies make the movement of capital easier and more rapid than ever before, they also provide the means by which to track — and therefore potentially regulate — these capital flows. Unless *legally* prohibited by treaty, new information technologies could be used by national governments to regulate or tax investment flows. Ironically, the MAI may be necessary to protect multinational corporations from themselves — from a system that they have designed to enhance their ability to extract digital profits.[34]

REORGANIZING THE WORLD OF WORK

The effects of the technological restructuring that enable the wonders of global digital information transfer are felt throughout the existing economy. As with all such restructuring, the most devastating results in North American society are found not in corporate boardrooms, but in the workplaces of millions of lower- and middle-class workers. The processes of automation, integration and networking that have taken place since the 1980s have resulted in the "massive erosion, deskilling and demeaning of work."[35] At the same time, the gap between the rich and the poor is increasing. Full-time secure jobs are dwindling. Former U.S. Labor Secretary Ray Marshall argues that we are currently facing the "worst economic crisis since the 1930's," due primarily to technological restructuring and international competition.[36] Prospects for improvement do not look good. The much-anticipated new jobs in the information and technology sector have failed to materialize on any large scale — certainly not to the extent needed to fill the gap created by technology-related job losses. In fact, it is likely that increased unemployment, underemployment, de-skilling and low wages will remain dominant trends in the future.[37]

It is important to understand that this process of technological restructuring is occurring in a labour system characterized by sexism and racism. Sexist and racist assumptions result in inequities in employment and contribute to how men, women and minorities are viewed in society.[38] In this context, job segregation has become a key way in which masculine privilege in the job market and patriarchy, in general, are maintained.[39] In fact, research has shown that the most desirable jobs in American society are also the least likely to employ females or African Americans.[40] It is even possible to suggest that jobs themselves have gender and racial identities. These identities can be determined by assessing a number of factors including the job's perceived complexity, the degree of autonomy, the level of authority over others, the supervision requirements, the degree of training, the salary, and the possibility of promotion. The jobs that rank highest in these characteristics, especially highly skilled, well paying jobs, are normatively "reserved" for white males. There is also ample evidence to suggest that as a higher proportion of women enter a field, wages fall and working conditions deteriorate.[41] While women have made some inroads into high technology management positions, particularly in small companies, there remain very few women in key decision-making roles. This is especially the case in the crucial areas of development, policy and production.[42] The United Nations has estimated that provided the current slow rate of improvement continues, women will achieve equality with men in the upper echelons of economic power in the year 2490.[43]

In view of this grim reality, it is not surprising that technological change in the workplace has never affected men and women in the same way.[44] In fact, a greater degree of occupational segregation occurs in high-technology industries than in other sectors and women are consistently underrepresented in these industries. Over time, the degree of occupational segregation is also more pronounced in sectors undergoing rapid technological change.[45] How does this occur? Research consistently shows that the gendered nature of technology has influenced labour markets by restructuring employment on the basis of each job's perceived intellectual content. As a result, new technologies provide opportunities to redefine jobs at the top of the hierarchy as male (requiring abstract or complex skills) and jobs at the bottom as female (requiring concrete or simple skills).[46] So what will be the likely effect of the new, high-status jobs created by new information and communications technologies in North America? Many predict that these few coveted "infojobs" will shift the balance of formerly "clerical" work opportunities from women to men. This will occur as new jobs increasingly require technical skills that women either do not have or that society defines as more appropriate for men. Meanwhile, jobs currently staffed by women are being deskilled and degraded by the introduction of more and more sophisticated technologies that perform increasingly complex tasks.[47] As a consequence, women's status in such jobs is being eroded and "women workers are bearing a disproportionate share of the burden of technological displacement."[48]

The underrepresentation of people of colour in the high status occupations and training institutions of high technology suggests that a similar dynamic occurs with respect to race. U.S. studies continue to indicate that job segregation by sex *and* race remains "the norm rather than the exception."[49] Unfortunately, there is still a lack of social science research that explores the effects of western technological change on the employment of women of colour in particular.[50] Notable exceptions, such as the work of historian Venus Green, Evelyn Nakano Glenn and Charles M. Tolbert II, however, do suggest that as the status of computer-related occupations has increased, the opportunities for African-American women in this sector have decreased. They also predict that future employment opportunities for women of colour in the West will largely be confined to unskilled work.[51] In addition, Green notes that while African-American women in the United States share a common experience of gender segregation with white women, they experience a secondary level of segregation as they are placed in jobs that subordinate them to white women as well as to men.[52] Not surprisingly, in view of this "double jeopardy," Black females represent "the most excluded group from valuable training opportunities."[53] Thus sexism and racism combine to create an even more intense form of economic subordination.

The way in which sexism and racism interact and are experienced by different women of colour is more complex than a simple secondary layer of discrimination. Such experience is complicated by different historical relationships to work. For example, for African-American women, this includes the legacy of slavery.[54] Recent research suggests that African-American women experience a form of "gendered racism" that may result in them being hired by white managers over African-American men because they are perceived as the more docile, less threatening of the two groups.[55] Once in these entry-level positions, however, upward mobility within the workplace is severely constrained by perceptions of incompetence based on both racist and sexist assumptions. Training and promotion opportunities are often denied as a result.[56] Such assumptions of incompetence are, ironically, accompanied by demands for job performance that exceed those placed on white men. When accompanied by sexual and racial harassment, neglect of specific needs such as childcare, and the stigma of affirmative action backlash, the work experience of many of these women becomes intolerable.[57] Far from alleviating these problems, the process of technological change in the workplace will likely exacerbate the situation. In an era in which, according to digital discourse, every sector of the economy will be altered by rapid technological change, the cumulative impact on women and women of colour could be devastating. Nevertheless, digital discourse continues to unproblematically sell the opportunities of the information age.

Another important trend in the current process of technological restructuring has been the increasing importance of "nonstandard" jobs. These types of jobs have been a major focus of feminist analysis relating to new information and communications technologies. This is not surprising because the majority of those workers being redirected into contract work, part-time work, telework and home work are women.[58] For women with caregiving responsibilities or mobility problems,[59] home work and telework, facilitated by the ability to transmit work online to the employer, have initially seemed to provide excellent opportunities — hence the appeal of the Compaq advertisement that sells laptop computers for use by working mothers. (See Figure 1.) However, further investigation indicates that the advantages of such arrangements lie almost entirely with the employer and a government in retreat from public expectations. Home work allows business to transfer the overhead costs typically incurred through the provision of a workplace to the employee who maintains a home office. In addition, supervisory staff can be replaced by electronic monitoring devices that record, for example, the number of keystrokes produced per minute. Alternatively, workers can be employed on a piecework basis, which removes all need for supervision. The government, meanwhile, is able to completely avoid the provision of public childcare or subsidies to private childcare facilities. Indeed, with women at home where they "traditionally" "belong," the dwindling welfare state can further divest itself of services that can be provided by women's unpaid labour in the home.[60]

The advantages of home work for women are not so clear. According to studies of women's work experience, home workers consistently "fare less well than their counterparts who went out to work to do a comparable job."[61] Not only does payment not take into account worker overhead, but it rarely offers benefits such as holiday pay, and does not provide any compensation for wide fluctuations in the availability of work. Work conditions and status are typically lower than for on-site jobs[62] and many workers express high stress levels due to the difficulties of juggling caregiving and work responsibilities, and the psychosocial effects of prolonged use of video display terminals (VDTs).[63] The high level of worker isolation endured by home workers is also problematic. In addition to reducing job satisfaction, this isolation prevents home workers from organizing trade unions to protect their rights. Alienated from one another and the predominantly male organized-labour movement, women home workers remain uniquely vulnerable to employer exploitation. Job loss is also a constant fear, as data entry positions become redundant due to technological change or are farmed out to cheaper off-shore locations. As a result, in Canada, many Pizza Pizza workers receive a paltry 25 cents for each order they handle and must actually pay their employers to rent their own computer.[64]

Considerable health issues are associated with computer work. Independent research conducted in the early 1990s outside the United States, and beyond the reach of the powerful computer industry, clearly links many health problems to the extremely low frequency electromagnetic fields produced by computer monitors. These health problems include various forms of cancer and increased rates of miscarriage. In fact, video display terminal (VDT) related illnesses have been called "the major occupational safety and health problem of the postindustrial society." As a result, medical experts have suggested a four hour limit on VDT daily work time.[65] Yet studies show that the many health risks associated with prolonged exposure to VDT radiation are not taken seriously by employers.[66] Even the very subject is becoming increasingly taboo in our computer-saturated society.

As technological restructuring creates decreased stability in the job market, workers are compelled to continually retrain themselves to meet the needs of technology and capital. The onus lies with the individual worker to meet the ever-changing needs of the market. Menzies describes this as part of a process of rising "credentialism," in which on-the-job experiential knowledge is becoming less important than accredited task-based and technology-specific knowledge.[67] Women's reduced access to technical training, systemic discrimination in the workplace with respect to such training, and the cultural perception of female technological incompetence places women in a disadvantaged position here, too.[68] Meanwhile, the provision of meaningful, low-cost access to training and equipment for groups traditionally alienated from technology remains inadequately addressed by either business or government.

None of these valid material or health concerns are addressed in the descriptions of technotopia that dominate digital discourse. Rather, like Compaq's advertisement, women are simply sold idealized images of themselves using laptops in the comfort of their own homes.[69]

RELEARNING WHAT WE ALREADY KNOW: THE LATEST GENDER CODE

The recognition of digital discourse as the offspring of the historical development of modernity and as surrounded by a context of globalized corporate capital certainly helps to demystify many of the elements that drive digital discourse and technology. Yet a crucial thread is missing from this history and context. Western women also need to understand how digital technology and discourse are implicated in the present-day construction of gender relations. The gendered implications of digital discourse go beyond the effects of the integration of digital technology into the technological restructuring of work in North American society, the exploitation of female labour in the South and increasing levels of commodification. As in all periods of widespread social and technological change, a process of cultural reconstruction is currently under way, and digital discourse plays an important role in that reconstruction.

For the purposes of this book's analysis, gender is to be perceived as a socially constructed category, a form of social difference that is embedded within complementary socioeconomic systems. This is not to deny the existence of distinguishable sexual differences — it is to point out that such differences have been heavily laden with cultural constructions that have served to elevate the status of traits deemed to be masculine over those deemed to be feminine. As noted in the Introduction, many social elements are gendered as either masculine or feminine, which creates a series of hegemonic dualisms. Despite the historical longevity of such dualisms, constructions of "masculine" and "feminine" are constantly being defined and redefined through the circulation of shifting ideologies and discourse. Assessing these past and present constructions of gender in discourse helps us to understand how relations of gender are created and perpetuated. Because these relations of gender play a very significant role in how we perceive ourselves and our world and help to determine what opportunities and limitations we experience in our lives, gender analysis offers some of the most significant insights about our society.[70] By combining gender analysis with a perspective that is woman centred and dedicated to ending oppression based on gender, race, class and sexual orientation, feminism at its best offers a unique, politically engaged methodology. The strength of such analysis lies not only in its ability to make an important set of power relations evident, but also in its ability to demystify, name and offer resistance to oppressive processes that we experience in our own lives.

Using such analysis, we can see that the dominant construction of femininity in the United States and Canada has undergone a number of significant characteristic shifts throughout history. Many of these centre on whether or not white women are expected to engage in paid work outside the home.[71] Women's work patterns have historically been manipulated to meet the particular needs of modern capital. As a pool of surplus labour, women are thus moved in and out of the labour force when required. For example, women were moved into munitions factories in the 1940s during World War II, returned to the home in the 1950s and then moved back to the workplace due to the expansion of the service sector in the 1970s.[72] Cultural discourses that rose and fell in each of these periods reflected the needs of capital, maintained patriarchal privilege and, to some extent, indicated women's resistance to prevailing gender stereotypes. *The Mary Tyler Moore Show* and *Rhoda*, for example, played an important role in normalizing middle-class white women's return to the workforce in the 1970s without threatening the full extent of masculine privilege or dominant race and class relations.

Since the 1980s, a process of widespread technological and economic restructuring is once again having profound effects on the lives of women. Worldwide, women continue to work more hours than men on average, rest less and perform a greater variety of tasks. While most of men's work is paid, the bulk of women's labour remains unpaid. Statistics suggest that things are getting worse. Between 1980 and 1994, women's share of the paid labour force actually *declined* by 2 percent or more internationally, to reach a mere 36 percent. At the same time, women constitute the largest and fastest growing share of the world's poor.[73] In the industrialized West, women are seeing many of the gains made by the women's movement to secure women's equality erode or disappear.[74] The "feminization of poverty" is becoming more and more common at the same time that so-called women's issues are being defined in increasingly narrow ways. The decline of the welfare state during this period has resulted in the contraction of government programs receptive to the needs of modern working women (that is, pay equity, childcare, assistance for single mothers) and a renewed reliance on women's unpaid labour as a "shock absorber" to cuts in social services. As a result of this regressive political climate and the effects of technological and economic changes that are reducing the total number of available jobs, many women are once again being forced back into "traditional" roles in the home. The phenomenon of home telework is an example of this. Women can be returned to the home, where they "belong," while shouldering not only the caregiving responsibilities for their families, but much of the financial burden as well. This movement of women's labour back in to the home also ensures that male privilege in the shrinking world of secure, paid employment is maintained.

Complementing these political and economic changes in the West is a resurgence in the rhetoric of the New Right that supports a return to traditional "family values." These values are accompanied by misogynist images of women in popular culture as either idealized "virgins" or vilified "whores." As Susan Faludi's *Backlash: The Undeclared War On American Women* painstakingly documents, the rise of antifeminist rhetoric in American popular culture, politics and society during the 1980s has left its mark on the society in which we live.[75] This process has continued through the 1990s, echoed even by the (apparently) more moderate eras of Bill Clinton, Tony Blair and Jean Chrétien.[76] Today, both television and movies provide a seemingly endless supply of violent, male-oriented dramas that rarely (if ever) feature an unproblematic independent woman. Television's latest idealized female characters, mainly found in comedies, feature the lonely and incomplete career woman *Ally McBeal*, the ditzy New Age sixties baby Dharma of *Dharma & Greg*, and the eternally well-coifed and directionless women of *Friends*. Meanwhile, *Baywatch*, a television show dedicated to the "careers" of bikini-clad lifeguards, continues to secure huge audiences worldwide.

In the emerging world of digital culture, the situation is not much different. After a decade of steady growth in women's enrollment in computer science programs between 1974 and 1984, the trend has reversed. Since 1984 a steady reduction in enrollment rates saw the proportion of women in computer science drop from its all time high of 37.2 percent to just 16 percent in 1996.[77] This period of decline coincides exactly with the rise of the computer industry as a dominant economic and social force in North America. At the same time, digital technology is being more and more vigorously defended as a field of masculine dominance through corporate design, use and advertising.[78] Assumed to be the purview of white upper-middle-class men, computer software and hardware has been designed and marketed with their interests in mind. It is not surprising, then, that there is a close relationship between the rise of many particular forms of technology and the pornography industry. As Carole J. Adams has perceptively noted, technologies that "take off" in modern western culture are often those that have a potentially lucrative early application; the high profits promised by these applications make it appealing to entrepreneurs. (Such an application is referred to as a "killer.app.") In the case of video technology, for example, pornographic videos drove the early development and proliferation of video technology. When no one else was investing in video, pornographers were quick to see its potential as a medium for their products. According to Adams, the lucrative application of CD-ROM technology is also pornography — interactive pornography that allows the "viewer" to choose his own "adventure." Likewise, the Internet provides a uniquely accessible and anonymous opportunity for the dispersal of pornographic material. It also allows users to become pornographers themselves, capable of disseminating their own porn over the Net and contacting others who share their particular interest.[79] There can be little doubt that the pornographic applications of the Internet, marketed almost entirely to heterosexual men, are at least partially responsible for its early growth. This also helps to explain how new information and communications technology has been constructed as a masculine domain.

The active exclusion of women from these latest technologies begins at a very early age: many of the computer games created for girls reproduce traditional gender and racial stereotypes with titles like "Barbie

Fashion Designer" and "Barbie As Rapunzel." In fact, Mattel Media's Barbie series has been very successful at tapping the growing "girl game" market.[80] A far cry from the action-oriented, violent fare created for boys, the newly developing market for girls games includes a variety of uninspiring "friendship adventures."[81] The majority of these reinforce existing sexist stereotypes by highlighting the importance of popularity and fashion for success. They also do little to help girls acquire computer skills or self-confidence.[82] While the variety of games available for girls has improved since 1997, the industry remains characterized by gender specific games.[83] Some teenage girls are clearly rejecting these "girl" games in favour of games designed with their brothers in mind.[84] Yet it is doubtful whether this represents a significant improvement. Such games merely engage in a reversal, forcing girls to take on stereotypical personas as violent warriors. At the same time, young girls (a demographic that has only recently been recognized as a particularly profitable one) continue to be inundated with the notion that "girl power" is intimately linked to their sexuality, popularity and ability to "have fun." Musical groups like The Spice Girls and television shows like *Buffy the Vampire Slayer* and *Sabrina the Teenage Witch* are examples of such cultural products. Packaged as a form of "postfeminism" — a kind of "feminism lite" designed for consumer culture — this discourse features highly individualistic, stereotypic and nonthreatening ideals of femininity that reinforce the regressive political and cultural climate.

"Free" from the feminist struggles of the past, women's "power" is contained within contemporary cultural discourses that reinforce very traditional views of women as individual caretakers, mothers or, most frequently, aggressive femme fatales. Feminism is associated with a distant past, with a form of so-called man-hating and radicalism that no longer resonates with the "new" woman of the 1990s.[85] Exceptional, successful businesswomen are held up as proof that feminism is no longer required, while legal, symbolic provisions of gender equality (as in the Canadian Charter of Rights and Freedoms) suggest that such equality *actually* exists. Unfortunately, as we have seen, the realities of many women's lives — whether in the public or the private sphere — simply do not support such claims.

Declining female welfare and the limitations placed on women (and men) by strict definitions of gender do not concern digital discourse.

Digital discourse is inscribed upon a background of a particular construction of self, society and the natural world that is white, western and masculine. As a result, stereotypes of white western masculinity — not femininity — feature heavily in digital discourse. Once again, we see the image of the lone white male subject, battling against natural and man-made elements and other individuals (often of a more "primitive" race), in order to promote a "universal truth" or himself. He is a man of reason in a world of chaos; a man in control of both himself and his world. However, in the hypermodern world of digital discourse, these elements are intensified and dramatized in the construction of the hypermacho man. More alone than ever before, it seems that the "hypermacho" man seeks to reassert masculine white privilege in an era of declining personal control over a rapidly changing world. He seeks to rebuild reality, entrench existing hierarchies, and escape the complexities of a diverse and unpredictable world. As a result, as we shall see, digital discourse is steeped in sexist language and racist assumptions.

This emerging construction of white hypermacho masculinity is not without precedent. Like constructions of femininity, constructions of masculinity have faced a series of significant changes in the recent history of modern North American society. Well-known feminist Barbara Ehrenreich argues in *The Hearts of Men: American Dreams and the Flight from Commitment* that constructions of masculinity in North America can be mapped historically in relation to their progressively noncommittal characteristics. Ehrenreich sees contemporary North American dominant masculinities as developing historically from a moral climate dictating responsibility, self-discipline and commitment to the role of breadwinner, to a climate endorsing "irresponsibility, self-indulgence and an isolationist detachment from the claims of others."[86] In many ways, her book anticipates the development of what is known as the Generation-X phenomenon, a masculinely gendered phenomenon that I would argue is the logical extension of the process of detachment and irresponsibility that Ehrenreich describes.[87] When extended into the hypermodern 1990s, and what theorists Arthur Kroker and Michael A. Weinstein refer to as the "will to virtuality,"[88] we can glimpse another manifestation of this same masculine model of technological, transcendental escape — enter, the hypermacho man. As we shall see, for this idealized man, digital discourse holds out the dream of another world; a cyberfrontier that appeals to long-enduring myths of masculine power and imperialistic control. Unfortunately, the accompanying emphasis on dehumanization and the flight from the body also diminish the value placed on the feminine, as reproductive relics from a primitive age. In a society where women's bodies are already degraded, further attempts to achieve a technologically mediated mind/circuit transcendence harmonize perfectly with a hegemonic culture of backlash that once again seeks to discipline feminine independence. Meanwhile, as meaning and knowledge are increasingly replaced by a commodified form of bit/information, feminist attempts to validate women's identity and experiential knowledge have lost much of their resonance and credibility in mainstream culture.

Chapter Two

BEYOND THE PACKAGING: WESTERN FEMINISM AND THE POLITICS OF ANTICIPATION

> The struggle to control the technologies of the new economy must begin with a struggle to control perception ...
>
> — Heather Menzies, *Whose Brave New World? The Information Highway and the New Economy*

AGAINST THIS BACKDROP of cultural backlash and a regressive political and economic climate, the technotopic promises of digital discourse appear even more dazzling and remarkable. Amid the intense, shiny packaging of this discourse, unfamiliar technical and theoretical language makes it difficult for many of us to determine exactly what is being promised, by whom and in exchange for what. Thankfully, there are many social critics and technology gurus only too willing to help us understand what it will mean to live in an increasingly digital world. While some of these writers and academics have offered important insights into the possible effects of current technological changes — the work of Heather Menzies, David Noble, Neil Postman and Langdon Winner come immediately to mind — the field is also plagued by a disease that has reached epidemic proportions in hypermodern society: technophilia.[1] Affecting both minds and bodies, technophilic views influence how we perceive our society and lend credibility to the advertising slogans of corporations that sell the new technologies of the digital age. Will new information and communications technology bring the "digital Nirvana" that technophilic social commentators would have us believe?

According to influential technophilic writers such as the McLuhan

Institute's Derrick de Kerckhove and the Massachusetts Institute of Technology's Nicholas Negroponte, the answer is a resounding yes.[2] Seeing the current developments in digital technology as both technologically and socially revolutionary and inevitably progressive, de Kerckhove and Negroponte largely ignore the history, socioeconomic context and inequitable power relations that underlie them. Technological determinism abounds in their work. For example, in one fell swoop, de Kerckhove both reifies technological change and prohibits any possibility of systematic social critique: "one could conceivably suppose that computerization uses business and government as the ideal milieu for growth and integration."[3] Revolutionary language and grand promises of a brighter future litter technophilic works. In the words of Negroponte: "The information superhighway may be mostly hype today, but it is an understatement about tomorrow. It will exist beyond people's wildest predictions ... we are bound to find new hope and dignity in places where very little existed before."[4] Similar forms of technological determinism and giddy enthusiasm are found in North American mainstream advertising and news media that tell us the information revolution will change our world and all our desires will be met in cyberspace. We'll be able to shop the world online. We'll have access to high-quality, limitless information and entertainment. We will do our banking at home — even work from home. Our children will discover new worlds of education. Virtual reality will allow us to experience new things in the safety of our homes. We will be able to express our views and a new era of democratic freedom will be realized. Such views of the future add a measure of credibility to the advertisements of telecommunications companies that dance across our television screens, vaguely promising to "connect" us to "the things that matter."[5]

In many ways, this "technovangelism" — whether found in advertisements, news media or in the work of technophilic social commentators — has become a convenient shorthand for promoting massive socioeconomic change in a way that limits the possibility of alternatives. As Menzies suggests, it provides a way to "control perception." Promises and speculation mask the complex social issues of inequality, oppression and the human and ecological costs of technological restructuring. Social change is conflated with technological change, which is viewed as a runaway train inevitably propelling human society "forward." But does it make sense to suggest that technology will solve sociopolitical problems to which it is intimately connected? Or that cyberspace will somehow escape the unequal power relations of the society that has populated and designed it? Could the emerging digital culture really represent an exception to and possible way out of the regressive popular culture and politics that women have experienced since the 1980s? Or is it more probable that this new digital age will be an extension of it?

WHAT ARE WESTERN FEMINISTS SAYING ABOUT THE DIGITAL AGE?

Western feminist scholars and writers, since the mid 1980s, have filled the gap created by many technophilic, mainstream approaches to new information and communications technologies. Their works draw attention to the gendered history of science and technology and to the low rates of female participation in the design, implementation and use of information technology. From their research, we know that the majority of Internet users are men,[6] that women are largely excluded from the development and administration of the etiquette for this new medium, and that, as a result, we must "make no mistake about it, the Internet is male territory."[7] Concern has been raised about the possibility of the Internet intensifying existing unequal power relations by creating classes of "information rich" and "information poor," along familiar lines of class, ethnicity, race, gender, language and region.[8] In addition, the significant role played by economic and military interests has been repeatedly acknowledged, in view of the Internet's origins in the U.S. Defense Department's ARPANET project.[9]

Not surprisingly, not all feminists have responded to or participated in this research in the same way. In fact, western feminists have understood the rise of new digital technologies in remarkably different ways. Currently, two approaches seem to have become popular and are widely circulated within academic women-and-technology discussions. These include what I identify as a more traditional *liberal feminist* view and an emerging *cyberfeminist* perspective, which is currently on the rise, particularly in North American universities. As we shall see, these perspectives seem to be a response to feminist approaches to technology (including many *ecofeminist* approaches), dominant in the 1970s and early 1980s, which are now problematized as technophobic and essentialist. These new approaches have also faced criticism because they raise other problems — especially as they relate to issues of class difference. Consequently, the usefulness of liberal and cyberfeminist approaches as vehicles for developing feminist political strategies that transform society remains questionable. In fact, it appears that much work remains to be done if we are to build a more inclusive theory, one that looks at the effects of technological change on women and takes into consideration the significance of how both *materially* and *discursively* constructed sexism, racism and classism affect women. What do these new feminist approaches offer?[10]

Liberal feminist approaches to new digital technologies focus on the existing obstacles to women's access to the Internet and technical training. The underlying assumption is that if more women were users and eventually became technicians, programmers and designers of computers and computer networks, the Internet's gendered nature would disappear over time. Some feminists even suggest that the perception that women's relationship with information technology is problematic is a generational one that is simply not experienced by younger women who have grown up with electronic communication.[11] The context in which such technologies are designed, implemented and disseminated, insist these feminists, is not inherently problematic. The main problem is simply that not enough women are accessing the Internet and reaping the benefits of online computer culture. In other words, in the field of digital technology, equal opportunity is the goal. As a result, feminists must work hard to get women into technology. The most compelling versions of this approach recognize a wide range of access issues, including not only the need for the greater availability of the necessary computer equipment, but also the need for gender-sensitive training opportunities, resource materials and institutional support. Liberal feminists also often attempt to respond to the development of regional disparities with respect to access. For example, groups such as the Alliance for Progressive Communication's (APC) Women's Networking Support Program is organizing global initiatives to ensure that barriers preventing women's access to the Internet in the developing countries of the South are addressed.[12]

These predominantly liberal feminist accounts are recommended by the pragmatism and common-sense logic of their approach. The words of Dale Spender in *Nattering on the Net: Women, Power and Cyberspace* sound eminently reasonable: "Given that I have to learn to live in the cyberworld, I want the best possible outcome that can be realized."[13] By advocating women's increased access to electronic communications and the creation of social policies that translate familiar notions of equality of opportunity into the context of virtual online communities, liberal feminists seek to attain this "best possible outcome" with unquestionably "do-able" strategies. The introduction and rapid spread of new information technologies is, for liberal feminists, no different from any other technological development; while women may initially lag behind in terms of access and may experience some cultural resistance to their participation, their presence online will eventually become unexceptional. The gender stereotypes that characterize the introduction of new technologies will ultimately disintegrate. Such an approach is sometimes supported by the use of historical evidence. For example, Spender uses the example of women drivers. Although the notion of women drivers was ridiculed in the automobile's early days, they are now unremarkable, and, in fact, statistics indicate that women today are safer drivers than men. While Spender recognizes that women remain "forever restricted to working with a product that men designed to fit men's lifestyles and hobbies,"[14] she argues that if enough women become involved in online culture quickly enough, they will presumably be able to influence the priorities and practices of "cybersociety." The trick is to act now — using strategies designed to meet women's specific access and educational needs — in order to ensure that women meet the challenges of yet another technological innovation.

Unfortunately, the liberal feminist "add women and stir" strategy remains unsatisfactory because it does not take into account the extent to which digital technologies are embedded in socioeconomic practice. It assumes that new digital technologies can somehow be divorced from the context in which they were designed, manufactured and disseminated. Like technophilic social commentators, liberal feminists seem to construct new technologies as neutral — as simple tools that are empty of any social, political or economic purpose or meaning. They also fail to recognize how these technologies are experienced very differently by different women and how they are implicated in larger processes of class, race and gender exploitation. In this way, liberal feminists *react* to the coming digital age, but offer little resistance to the interests that propel it. They see new digital technologies as just another technological development to which women must adapt. We have no choice if we want to be involved in determining how the new digital economy will be constructed, or at least given the opportunity to compete within it.

But what if the Internet does not resemble other technological developments? According to emerging cyberfeminist perspectives, the world of new digital technologies is a whole new ball game. One that women will win. Cyberfeminist hopes for the social effects of the Internet are far more expansive and optimistic than most liberal feminist approaches. Echoing the rhetoric of many technophilic social commentators, cyberfeminists argue that the Internet is unlike any previous technology.[15] They see computer networking and the development of online communities as technological manifestations of our postmodern times. In fact, Susan Myburgh contends that the Internet presents opportunities for political engagement the likes of which western industrial society has never seen before. Because the Internet is an interactive form of media that potentially invites a diversity of users to communicate in a new world that ignores one's gender, race, class, religion and age, Myburgh argues that "it is impossible to *do* hierarchy in cyberspace."[16] By adopting diverse online personalities with different or unspecified genders, members of online communities can ostensibly free themselves from the constraints of heterosexist norms. Minority voices become mixed with those of the majority. Identities become multiple and simultaneous and disembodied personalities subvert traditional categories and stereotypes. According to cyberfeminist Sadie Plant, women are uniquely suited to these new online environments. In fact, the "computers and the networks they compose run on lines quite alien to those which once kept women in the home" because the new information economy reduces the significance of physical strength to the world of work. In its place is a demand for "speed, intelligence and transferable, interpersonal and communications skills" — skills that Plant argues women have in spades.[17] Women are also well versed in role-playing and imitation, which makes them good candidates for the virtual worlds of the future. In addition, women's ability to adapt to changes in the workplace makes them ideally suited to the emerging world of part-time, discontinuous work.

Plant's cyberfeminism is centred on the theory that women share a great deal with both nature and machinery: all three are manipulated for the benefit of men. She argues that new technologies will destabilize masculine dominance by reducing men to the status of mere "users" in the face of a cybernetic world that is nonlinear, self-replicating, self-organizing and self-designing. The machines of man, she suggests, are evolving beyond men's control. As a result, Plant contends that new technologies, far from reasserting masculine supremacy, will actually represent "the return of the feminine, perhaps even the revenge of nature." The "posthuman world" that Plant suggests we are headed toward as a result of cybernetics, is a world of female cyborgs. Cyberfeminism will replace patriarchy. Cyberfeminism is "simply the acknowledgement that patriarchy is doomed" by its own technologies. It is therefore "not a political project, and has neither theory nor practice, no goals and no principles."[18]

One of the undeniable strengths of cyberfeminism is that it presents women with an optimistic alternative to theoretical positions that seem to relegate women to the status of victims within a context of a vilified, all-encompassing patriarchy. By affirming women's abilities and strengths, cyberfeminist perspectives resist replicating patriarchal constructions of women as technologically incompetent beings who are exiled from the world of technology. Such views are influenced by post-structuralist theories and, especially, by Donna Haraway's widely cited "A Manifesto for Cyborgs."[19] The latter is an attempt to move beyond traditional dualisms that associate women with nature and men with culture and technology by introducing the concept of the cyborg. The concept is very seductive because it seems to offer a theoretical way out of common western patriarchal understandings of the world. It breaks down the division between the artificial and the natural by arguing that this distinction is no longer practical in modern technological society. We are neither nature nor culture: instead we are cyborgs, part organic and part technological. Cyberfeminists have adapted this concept and used it to celebrate the liberatory potential of new technologies for women. Rather than identifying men with technology and women with nature, Plant emphasizes the link between women and technology: "Machines and women have at least one thing in common: they are not men."[20]

The emergence of cyberfeminist online groups such as geekgirl, Cybergrrl, Nerdgrrl and so on reinforces the optimism of cyberfeminist theory by providing "real-life" examples of how women are using the Internet for potentially subversive purposes. In fact, if we examine alternative feminist online cultures in isolation, cyberfeminism does have a certain resonance. Some women are using these new technologies for positive ends, networking with others, creating virtual communities, finding creative outlets. They also rhetorically see their activities as political — as subverting existing gender constructs by proving that women can use technology for their own purposes. But is it enough to look at these alternative online cultures in isolation? Can the surrounding historical and social context be so easily nullified with a flip of the modem switch? Is cyberfeminism ultimately realistic and politically helpful?

There can be little doubt that cyberfeminism's theoretical goal is a worthy one: it does seem counterproductive for feminists to perpetuate the same gender hierarchies in our work that we seek to overcome in mainstream culture. If we ever want to overcome sexism, we must also overcome stereotypes that exclusively associate women with nature and men with culture. And, quite arguably, ecofeminist perspectives have often been guilty of replicating patriarchal views by asserting that women have a unique relationship with nature.[21] They have also frequently glorified a distant, matriarchal, goddess-centred past that seems remote for many of us in the modern world. Uncomfortably similar to conservative ideologies, ecofeminist perspectives do seem to perpetuate the naturalistic view that men are the holders of power and women are the guardians of morality. Yet by so distancing our theory from existing women/nature and men/culture dualisms, we may fail to recognize their enduring presence and effects in our society. If we acknowledge the role that patriarchal dualisms play in how we experience our lives, are we necessarily doomed to perpetuate them? I don't think so. Understanding that women have been historically associated with nature in western culture does not mean that women are necessarily more in tune with nature or that we must embrace patriarchal dualisms as universal, essential or "real." It simply means that they have played a role in how unequal power relations are perpetuated in our society and continue to do so. It means they are a cultural problem that both men and women need to address. In any case, it remains questionable whether cyberfeminism actually is a theory that goes beyond traditional binaries. By so intimately linking the feminine with the technological, could cyberfeminists simply be reversing patriarchal dualisms? How ethical is it to replace the existing binary with a new one that places women, nature and technology on one side and men on the other?

Cyberfeminism also raises important questions about the extent to which new digital technologies are novel and unprecedented. Sherry Turkle, MIT professor and pioneering researcher of people's relationships to computers, agrees that the Internet allows us to "develop models of psychological well-being that are in a meaningful sense, postmodern."[22] Her recent book *Life on the Screen: Identity in the Age of the Internet*, explores the significance of the Internet to personal identity.[23] Turkle argues that the Internet, in fact, does encourage identities that are multiple, decentralized and flexible. Yet the data she examines raise the question of whether gender norms are actually being subverted by "gender-swapping" or simply repeated with ever-increasing frequency as bad reproductions. As Turkle notes, gender-swapping is not easily achieved. It demands an ability to speak, behave, *act* like the other sex. While such an act may lead to a greater recognition of how gender infuses and disciplines our use of language and behaviour, it may also strengthen existing stereotypes as they are more frequently used as shortcuts to establishing the desired gender identity. Certainly, playing with gender identity is nothing new: from Shakespeare's *As You Like It* to *Victor/Victoria* to *Tootsie*, gender-swapping and cross-dressing have had an enduring presence in the entertainment of western culture. Still, it is difficult to imagine anyone suggesting that by dressing in drag the comedy group Kids in the Hall destabilized gender myths and encouraged more flexible identities. Nevertheless, online gender-swapping seems to be expected to do just that. Despite prevailing beliefs in popular culture about the radical new world of "cybersex," research indicates that "gender roles tend for the most part to remain stereotypical" in such environments.[24]

In fact, the more we compare cyberfeminist theory to actual

women's experiences online, the more difficult it is to see the digital age as conducive to feminist social transformation. If women are, as Plant suggests, so uniquely suited to cyberspace, why are so few online? And why, when they are online, do women participate less?[25] Analyses of women's online experiences, which often include gender-based harassment, exclusion and even stalking, also indicate that cyberspace is far from subverting existing power relations.[26] In fact, it is often because of sexual harassment that many women choose to assume a male gender identity while online. In the words of feminist researcher Larissa Silver, far from a radically different scenario, the world of new digital technologies can be more accurately described as "same message, different medium."[27] We must be wary of exaggerated claims about the Internet's novelty and its ability to technologically circumvent sexist stereotypes and discourses.

Cyberfeminism becomes even more precarious when we explore the social and economic context that surrounds new technologies — a context characterized by ongoing sex, race and class inequalities that do not simply "disappear" on the other side of the screen. Plant's conceptualization of women's ability to adapt to part-time, discontinuous work as representing an advantage in the new digital age is deeply problematic. Noting in passing that in 1992 women still earned a mere seventy-five cents to every male dollar, Plant suggests that women's sights are now set "beyond traditional focal points," leading to the growth of many successful women-owned-and-operated businesses.[28] Yet the extent to which the majority of women's experiences of technological restructuring is *not* characterized by "choice" must not be ignored. The individual successes of particular privileged women does not make the realities of women's economic subordination disappear. Nor does it negate the fact that women still face huge obstacles of access and literacy; women still represent two out of every three illiterate people in the world.[29] Insufficient time and money are also major inhibiting factors to women's use of digital technologies.[30] While affluent western feminists may see themselves as "cyborgs" as they use digital technologies for creative and professional purposes, less advantaged women — such as those who assemble computer equipment or enter data — experience "cyborg" life in a profoundly different and exploitative way. Plant's analysis is creatively and optimistically presented. But it remains a difficult one to reconcile with the present-day realities of work and education, many of which continue to disadvantage women, women of colour and people of colour. Material exploitation, class difference and historically and discursively constructed oppression are not recognized as problems that must be addressed in cyberfeminist theory. This is perhaps not surprising; cyberfeminism is, after all, not a "political project" but an "acknowledgement" of an inevitable, technologically determined future. Thus cyberfeminism's rhetoric of an idealized cybernetic future conceals the power relations of race, class and gender.

Fortunately, the significance of existing power relations continues to be acknowledged and problematized by many other feminists who have been influenced by postmodern theory. For many, this has involved exploring the role of discourse in perpetuating inequalities.[31] The growing recognition of scholars and social commentators of the significance of language may be understood as fundamentally connected to an economic process: the shift in the balance in economic life from production to consumption, and from manufacturing industries to service, culture and leisure industries. As a result, social theorists now appreciate more and more that key areas of social life are becoming increasingly centred on the media, especially television, but also on other forms of popular culture. The extent to which all aspects of our lives are commodified in hypermodern society through the cultural productions of capitalism must not be ignored. As critical linguist Norman Fairclough notes, language "has become perhaps the primary medium of power and control" in the modern industrialized world.[32] We are not only being sold products through popular culture, we are also sold lifestyles, identities, fantasies, even our "selves." And, as members of the paid or unpaid workforce, we are being disciplined and subtly (or not so subtly) disempowered. Menzies refers to an analogous process in corporate technological education programs as "training for compliance."[33] Since it is intrinsically important to understand how language functions through discourse, to understand language may also be necessary in any contemporary political strategy.

As a result of feminists' growing recognition of the importance of discourse, many have begun to use *gender discourse analysis* to understand how power is circulated in our society. The political implications of gender discourse analysis are significant because such study recognizes language as a reflection of existing unequal gender relations and also as an important component in how gender is accomplished. The gendered nature of many current technological discourses is recognized by Claudia Springer's *Electronic Eros* and Anne Balsamo's *Technologies of the Gendered Body*.[34] Both works provide thoughtful analyses of how elements of popular culture, and, in the case of Balsamo, institutional practices, serve to structure the way in which we experience ourselves and our bodies in a technologically mediated world. Neither traditionally liberal nor cyberfeminist, these feminists offer a more compelling approach to understanding the digital age through discourse.

Springer's work deals exclusively with fictional representations of cyborgs to demonstrate how techno-erotic imagery expresses "contemporary cultural conflicts over sexuality and gender roles."[35] Her examination of the contemporary eroticization of technology leads her to argue that gender constructions in western culture are in a state of flux. She suggests that the dispute over gender is intensified by the growing cyborg phenomenon in popular culture. This phenomenon, she contends, highlights anxieties that the future of human beings, in general, may also be in dispute. According to Springer, many of popular culture's cyborg narratives reproduce traditional patriarchal constructions of gender. That is, far from putting forward a sort of posthuman egalitarianism, feminism or androgyny, sexist stereotypes endure in cyborg narratives. Yet Springer still argues that it is possible to conceive of a future in which "technology will become part of an egalitarian social configuration."[36] In any case, Springer agrees with the view that "it may be too late to reject the cyborg existence. We are already jacked in."

Whether or not we are skeptical of Springer's optimistic conclusion, her work does provide a useful analysis of gender imagery in our technologically oriented society. Her insights are reinforced by Balsamo's analysis, which examines a wider array of discourses and practices, including not only popular culture's cyborg narratives and films but also feminist bodybuilding, cosmetic surgery, pregnancy and new reproductive technologies, and virtual reality. Balsamo demonstrates how these technologies are shaped by male gender interests and often reinforce traditional power relations between men and women. Together, Springer and Balsamo make it clear that western culture continues to perpetuate a very gendered view of what technology is and who men and women are in relation to it. Yet something is missing from both analyses. Although Balsamo insists that gender is not simply a matter of representation and discourse but is also the effect of social, economic and institutional relations of power, she admits that she only "implicitly" addresses the question of the relationship between discourse and the material conditions of women's lives.[37]

It is necessary to do much more than this. These "material conditions of women's lives" are important. As communications professor Carole Stabile argues in her exceptional book *Feminism and the Technological Fix*, feminist practice necessitates that it is not enough to interpret or "read" our own cultural reality. In addition, we must recognize that "there is little that is sexy, intellectually exciting, or conducive to postmodern aesthetics about the gray and dull realities of economic disadvantage." While, it may seem "more charming to pursue the trope of the cyborg winging her way through the gossamer realm of the postmodern," this kind of analysis fails to recognize the extent to which all such future discourses are being sold "on consignment from capitalism."[38] We must, therefore, remain ever aware of the concrete causes and effects of discourse. It is difficult to deny that economic interests are, at the very least, implicated in the production and circulation of discourse. In other words, we are rarely if ever completely disinterested speakers.

If discourse is so intimately related to economic conditions, why is class analysis often missing from academic feminist theory in the field of technology and social change? Stabile suggests that feminist perspectives have been influenced by the class context in which they are produced. In a climate where western feminist theory largely arises from the academy, Stabile sees the feminist neglect of class as an inability on the part of many western feminists to recognize their own class privilege. The limitations of these theories for widespread political practice may therefore be linked to the institutionalization, professionalization and, ultimately, the commodification of western feminist theory. It has been widely acknowledged (and sometimes lamented), in recent years, that western feminism, in its academic incarnation, has focused less on the material and sociological than on the symbolic and representational. This may also help to explain why much academic feminist theory makes little effort to politically engage with or communicate to a nonacademic audience. It has also frequently represented only the specific interests of upper-middle-class white women. As a result, the fact that technology affects different women differently — across class and race lines — is often not considered.[39] The challenge of creating feminist theory that is both politically relevant to material struggles faced by diverse women and avoids universalizing women's experience is perhaps the most central issue facing western feminists today.

Thankfully, the academic study of technology and social change does not entirely lack research that explores the socioeconomic implications of new technologies. Heather Menzies, whose work focuses on the contemporary phenomenon of technological restructuring in the workforce, has done the critical study of technology an invaluable service by addressing these issues. In *Whose Brave New World? The Information Highway and the New Economy*, she recognizes technological restructuring as the main cause of the recent social spending cuts and deficit reduction in Canada. Using both statistical data and case studies, Menzies shows how automation, computerization and networking has benefited a privileged minority of big business interests at the expense of the Canadian workforce. She also deconstructs discourses that prevailed in the 1970s, 1980s and 1990s and demonstrates how these discourses, ranging from technological change to free trade to deficit reduction, have defined the parameters of social debate. Throughout her analysis, she highlights the disadvantages women face: women's reduced access to technical education and training; the recent increase of the wage gap between men and women; the fact that since the 1970s and 1980s women have worked as hard or harder than in previous decades but for less compensation. Menzies even notes the fact that women are disproportionately represented in the ranks of contingent and low-end communications jobs and focuses on recent alarming data that indicate women are more likely to be subject to electronic monitoring on the job than men.

Despite these significant insights and the recognition that technological restructuring is not simply an economic issue but also a sociopolitical issue that "affects our language, consciousness and identity," Menzies does not integrate gender into her theoretical framework. Gender is relevant in her analysis of the *effects* of technological restructuring but is not discussed as a significant aspect in its *cause*. As a result, while Menzies provides an excellent study of how technological change today affects women, she does not offer a feminist analysis of why it does so.[40] In this way, Menzies mystifies the link between patriarchal discourse and practice. This is a problem. How can we understand the gendered effects of technological change without recognizing the context and structure of technological discourse as patriarchal? Alternatively, how can we account for the endurance of racial discrimination without recognizing the existence of systemic and discursively hegemonic racism?

Toward a Feminist Politics of Anticipation

In light of the many important insights raised by current feminist approaches to the emerging digital age, I use gender discourse analysis in this book as a way of developing a larger sociopolitical project that I call a *feminist politics of anticipation.* At the heart of this approach is the view that hypermodern culture requires critical tools that allow us to keep pace with the emerging trends of our high-speed society and that will help us to determine viable, active strategies to influence positive change. If we are to avoid forever responding to realities that are already in decline or are being rendered unrecognizable by contemporary developments, we must learn to *anticipate* social and technological change. For this reason the feminist politics of anticipation that I advocate begins with gender discourse analysis. Emerging discourses offer clues about how our society is developing — they let us see how knowledge is constructed, how truths are deployed and identities altered. They also offer us a way to see how power is being circulated and how different social interests would like to organize our society in the future. Gender discourse analysis presents politically engaged feminists and other social justice activists with a unique opportunity to *anticipate* and critically respond to the emerging digital culture before it becomes widely accepted.[41]

There is more to a feminist politics of anticipation than gender discourse analysis, however. One must also link the discourse surrounding new technologies with the overall socioeconomic context in which these technologies are situated. The interrelationship between discourse and material conditions must be continually acknowledged and explored. This means overturning the schoolyard adage that "sticks and stones may break my bones, but words can never hurt me." Words actually do hurt. It is through language that our world is socially constructed and given meaning and that power relations are formally established and reproduced. Recognizing that words have consequences has been crucial to the recent development of discourse analysis as a field that engages in the critical study of text and talk. If discourse is indeed the battleground on which sexist power relations are produced and reproduced, then critical feminist engagement with discourse becomes a powerful form of political action. This is indicated by our discussion of how dominant ideas about women and minorities affect their employment experience. It is not enough simply to read culture. A feminist politics of anticipation must use this reading in order to understand concrete political and economic realities. Just as words have consequences, they also have causes that are rooted in material realities which we must also understand.

Although gender discourse analysis is not the only element of an anticipatory politics, it is one of the most crucial and one of the most challenging. In many cases, it can also be intimidating. Exploring digital discourse is difficult because it involves entering a world that is neither welcoming nor familiar to many women — a world that, because of its intimate association with technology, is culturally defined as a masculine sphere of influence. Women's exclusion from particular discourses is a familiar theme that affects our experience of popular culture and influences the direction of scholarly research. As a result, so far, feminist media theory and cultural studies have most often focused on genres popular with female audiences, such as soap operas, romance novels or women's magazines. With the exception of pornographic materials and science fiction, little feminist research of cultural products designed primarily for male consumption has been conducted. The field of computer- and telecommunications-related material has been particularly neglected due in part to the sense that "to talk about women and technology in the same breath seems strange or even incongruous."[42] While feminists are aware of women's exclusion from technology-related sectors, it nevertheless may subconsciously create an aversion to engaging with this field. For example, we may feel unqualified to assess a discourse that uses jargon we are not completely familiar with.[43] In addition, new materials associated with the "cutting edge" may seem less attractive to feminist scholars due to the perception that, like pornography, they do not represent the norm but rather an extreme exception. Yet just as feminist analysis of pornography has yielded important insights into the dominant myths of masculinity that pervade our culture, so does the analysis of new discourses provide important clues about shifting masculinities in an increasingly digital world. In addition, for many, gender study has simply meant study in which women are the subject; the fact that men too have a gender and are affected by constructions of masculinity and femininity is sometimes overlooked. Instead of avoiding the seemingly restricted space of digital culture, it is important to travel deep into it, to learn the dialects spoken there and to explore its most disturbing and seductive elements. Only then can we begin to crack the gender code it creates and perpetuates.

This book analyses the discourse of *Wired* magazine from 1993 to 1998.[44] *Wired* is an excellent example of the new ideological discourse that is emerging around new information and communications technologies. As well, it is a magazine that has become an icon of digital culture. As such, it provides a backdrop against which we can reflect on the dominant ideas that are central to much of digital culture and that are beginning to be expressed throughout North American society in general. These ideas have very real material causes and consequences. The discourse analysis I use is influenced not so much by postmodern theory's attempt to interpret a postmodern world as by a historical, material, feminist methodology, which uses gender discourse analysis in the political struggle against gender, race and class oppression. It is influenced by the feminist scholars and writers discussed in this chapter and draws substantially on the theory and case studies of such discourse analysts as Teun van Dijk, Norman Fairclough and Roger Fowler; cultural theorist Stuart Hall; feminist linguists Deborah Cameron, Julia Penelope and Susan Ehrlich; and feminist media scholar Liesbet van Zoonen.[45] Knitted together, the insights of these scholars provide the methodological grounding for our exploration of the *Wired* world.

My analysis of *Wired* will also draw upon the framework provided in Chapter 1 to interpret and situate the ideas, beliefs and practices that are put forward through digital discourse and to reveal the digital ideology that fuels it. The emerging digital discourse in *Wired* is analysed in its larger socioeconomic and cultural context. It is also crucial to understand *how* digital discourse expresses meaning and what discursive tools it uses to do this. This will help us to identify its influence elsewhere in popular culture and to determine how our own discourses can be framed to resist and counteract it. In addition, we need to explore grammatical structures (for example, how words and sentences are structured, punctuated and ordered) and dominant themes and topics; the significance of style (what words are chosen, what they invoke, the tone of the overall text); and alliteration, rhyme, irony, metaphor and hyperbole in order to understand how digital discourse attempts to persuade and motivate others.[46]

In hypermodern culture, it is important, too, to recognize the significance visual representations and graphics have in the meaning and method of digital discourse. The balance of communication in the modern television era has tilted away from words and numbers toward pictures and images. This change, said to be "accelerating and increasingly computer-led,"[47] is exemplified by the many high-quality computer-generated graphics and by the considerable attention the editors of *Wired* pay to the magazine's visual impact. This "multi-modal" analysis of *Wired*, therefore, explores text and graphics as two interdependent components of the same discourse.[48]

TIME TO GET CRACKING ...

According to *Wired*'s recent handbook about English usage in the digital age, a "cracker" is "an intruder; one who breaks into computer systems, 'cracking' them."[49] We must now become "crackers" so that we can understand the forces that presume to construct our future. We must enter this new *Wired* world armed with the critical tools of feminist discourse analysis. And, as "crackers" of a system that exists beyond the computer, meet the digital generation.

Chapter Three

THE 'WIRED' MACHINE

There are a lot of magazines about technology.
Wired is not one of them. *Wired* is about the most powerful people on the planet today — the Digital Generation. These are the people who not only foresaw how the merger of computers, telecommunications and the media is transforming life at the cusp of the new millennium, they are making it happen.
Our first instruction to our writers: Amaze us.
Our second: We know a lot about digital technology, and we are bored with it. Tell us something we've never heard before, in a way we've never seen before.
If it challenges our assumptions, so much the better.
So why now? Why *Wired*? Because in the age of information overload, the ultimate luxury is meaning and context.
Or put another way, if you're looking for the soul of our new society in wild metamorphosis, our advice is simple. Get *Wired.*

— Louis Rossetto, "Why *Wired*?"
Wired, January 1993

WITH THESE WORDS, *Wired* magazine proclaimed its presence as a North American periodical with a difference; a magazine dedicated to describing and understanding the newly formed "digital generation" and to providing the "meaning and context" behind the latest technological revolution. Dubbed "the change leaders of our time," *Wired*'s "digital generation" is intimately connected to the rise of new information and communications technologies and to the emergence of digital discourse as a significant force in western culture and society. This generation is composed of the same computer and telecommunications industry elites that are at the forefront of new media technologies and that are the primary proponents of the race to get online — or in *Wired*'s words, to "get wired." Because of this, *Wired* would seem the perfect medium through which those of us who find technological change daunting could gain access to the high-tech world and learn to understand the "soul of our new society."

In Canada alone, 48 percent of workers use computers in the workplace[1] and are therefore immediately affected by developments in the computer and telecommunications industries. Indeed, as many workers are painfully aware, technological change in these industries can result in the restructuring of their jobs or in their work being declared "redundant." Those who are not affected directly through the workplace find that other aspects of their lives are influenced by a technological revolution that promises to change everything — from preferred modes of entertainment to conceptions of time and space in a world of e-mail and digital cell phones. Parents have seen computers infiltrate the classroom with a growing intensity as teachers attempt to prepare the next generation for a new information age. *Wired*, then, should speak to all of us — even those who are not particularly interested in "surfing the Web" — because we all have an interest in understanding "the most powerful people on the planet." This may be especially true for women, who find themselves most dramatically affected by these new technologies that facilitate the growth of disadvantageous work forms, like home telework, and that rely upon exploiting female labour in the South. Indeed, in view of women's exclusion from technology historically, the current revolution in digital technology could be expected to raise numerous concerns for women. Yet, according to Paulina Borsook, a former contributor to *Wired*, the magazine elicits strong negative reactions from women who have turned to its pages for insight and entertainment:

> When thirty women at a San Francisco talk about women and the Internet (a pretty self-selected bunch relatively comfortable with technology) were asked if they read *Wired*, they all raised their hands. And when they were asked if they hated *Wired*, they all raised their hands. It appears women, even the techno-initiated, generally do not like *Wired*.[2]

There are many reasons for this. As we shall see, the magazine is definitely not designed for female readers. According to the 1997

Subscriber Study, conducted by Beta Research, *Wired* caters to a very exclusive audience. The average *Wired* reader is a thirty-nine-year-old male college graduate with an annual employment income of US$83,000 and a household income of US$122,600. He is most likely to live in the United States and to work in a professional or a managerial position, but also may work in high executive positions or be self-employed. Interestingly, *Wired* provides no data about the ethnic or racial background of its readership. In response to a request for demographic information from *Wired*, I received a brief e-mail from a customer service representative who informed me rather bluntly that the readership consists of "mostly well-educated white guys. Surprise!"[3] Whether or not this is the case (and it probably is), the assumption of a white readership, despite the lack of quantitative data to support the claim, indicates the tacit exclusivity of *Wired*'s readership and the presumed audience for which the magazine is designed.[4]

The *Wired* reader not only embraces the new information age but also sets about making it happen through his professional endeavours and private consumption. According to the 1996 Intelliquest V3.0/Business Study, *Wired* magazine readers are more "Wired to the Net" than those of such competing publications as *NetGuide*, *Internet World*, *MacWorld*, *PC/Computing* and *PC Magazine*. This statistic indicates a high degree of interest and participation in online communication and information services, which are, according to digital discourse, crucially important to a technotopic future. To potential advertisers, the *Wired* reader is likely to be an eager consumer of new digital technologies, both at work and at home. In fact, 38.2 percent of *Wired* readers hold high-ranking managerial positions in which they are likely to make important decisions about their company's information systems. According to the separate "Top Management Business Influencer Profile" compiled by *Wired* about this elite group of readers, these individuals are slightly older than the magazine average (forty-two) and have an even higher personal income of US$115,700 and household income of US$151,500. In other words, these readers are very similar to the "digital generation" that *Wired* identifies as "the most powerful people on the planet" — those who make the decisions that will affect the lives of millions. In reading *Wired*, the digital generation reads itself to some extent.

And apparently they like what they read. More than 83 percent of readers have read three or four out of the last four issues and they have also picked up each issue an average of 5.59 times. As well, survey results indicate that more than 70 percent of readers read *Wired* during their leisure time, in the privacy of their own home, and that all readers spend an average of two hours reading each issue. The 1997 Subscriber Study found that more than 90 percent of readers take some form of action after reading an issue. For example, upward of 62 percent mark or clip items in the magazine. Response to advertisements in *Wired* are reportedly very high. Over 77 percent of readers surveyed indicate that they respond in some way to advertisements in the magazine. The majority visit an advertiser's Web site and 22 percent say that they have purchased an advertised product. Certain issues of *Wired* are also coveted by readers; just over 66 percent of readers save past issues for future reference, while 28 percent pass issues on to friends and relatives. Such "sharing" ensures that each copy of *Wired* is read by a total of 3.3 readers.[5] These statistics indicate that contrary to those female readers polled by Paulina Borsook, *Wired*'s male readership enjoys, supports and is influenced by the magazine.

If *Wired* is an exclusively male-oriented magazine, well loved or not, why should women care about what it has to say? Because what *Wired* says does, to some extent, matter. As technological restructuring continues to affect almost every aspect of our lives and, as we have seen, disadvantages some of us more so than others, we must grasp more than just the gems of wisdom provided by corporate slogans like Clearnet's "the future is friendly." In order to see what lies behind such a seemingly benign message, we must lift the veil of technical jargon and explore how those at the forefront of these changes are constructing the future and what is at stake in the delivery of these messages. In other words, what is behind these all-too-familiar corporate slogans? The pages of *Wired* provide the digital generation with a forum where it can speak in its least apologetic tones. At home in its world of technologically mediated graphics and hypermodern design, and surrounded by the "friendly" faces of technologists, industry moguls and technophilic social commentators, *Wired*'s digital generation is free to speak its collective mind. *Wired* has made itself an important cultural symbol for digital elites and seeks to put forward their vision of the future. Thus *Wired* provides a medium for the delivery of those messages associated with a significant strain of elite digital discourse and ideology — one that has become increasingly influential in mainstream society.

To understand *Wired*'s message and why it should matter to women, we must understand *Wired* as a medium. We must determine where this discourse is coming from and where it finds its sources of socioeconomic power. Who are the creators of *Wired*? What are their positions — and what is the position of the magazine itself — in North American society? What interests do the publishers of *Wired* support? How powerful is this emerging discourse, and how has it come to enjoy its current status?

AND THEN THERE WAS *WIRED* ...

The first issue of the San Francisco–based *Wired* magazine appeared in January 1993. At that time, the realities of the Internet were even less commonplace in the daily lives of most North Americans than they are today. Although three years had passed since the September 1990 issue of the cyberpunk magazine *Mondo 2000* proclaimed that "The Rush is On!" and that the process of colonizing cyberspace had begun,[6] television information programs had not yet added e-mail and Web site addresses to the end of every episode; and Al Gore had not yet made his famous speech at the 1994 International Telecommunications Union Conference in which he espoused the glories of an information system that will "make possible a global information marketplace."[7] In fact, it wasn't until a year after *Wired*'s first issue that the information superhighway became a favourite topic in the mass media.[8] Meanwhile, on the West Coast of the United States, *Wired*'s cofounders, Louis Rossetto and Jane Metcalfe, had identified a potential market for a new magazine that would catch the wave of new information and communications technology and both reflect and influence the development of the new information society that it facilitated. With an initial capital investment of $75,000 from Nicholas Negroponte, who later became the magazine's only senior columnist, *Wired* was born.

Since its first issue, *Wired* has experienced phenomenal commercial and critical success, suggesting that it has tapped into a lucrative market and is to some degree, original. Even in its first year *Wired* received a number of important awards, including *Ad Week* magazine's Startup of the Year Award for 1993, and *Advertising Age*'s Best of 1993 award. In 1994 and 1997, it received the American Society of Magazine Editors' National Magazine Award for General Excellence, and was nominated for this award in the intervening years. By 1996, *Wired* had made its first appearance on *Ad Week*'s list of the Ten Hottest Magazines of the year, at number five, ahead of *GQ* and *Entertainment Weekly*. According to *Ad Week*'s Adam Shell, in the space of one year, from 1995 to 1996, *Wired* had almost doubled its advertising revenue.[9] Equally if not more important for the magazine's survival, however, has been *Wired*'s ability to maintain an extremely healthy rate of circulation growth. According to marketing information compiled by the Audit Bureau of Circulations and published by *Wired*, the magazine grew by an astonishing 346 percent between January of 1993 and August 1995.[10] While this astonishing rate of growth could not be sustained, *Wired* continues to enjoy good health, with advertising up 9.6 percent in 1996, and circulation up 32.6 percent from 1995 figures.[11] According to the latest figures available from the Audit Bureau of Circulations, total single-copy and subscription sales have continued to grow through to June 30, 1997, at which point paid subscriptions totalled 347,465 copies.[12]

It is important to note that *Wired*'s success has not been achieved in a period of general industry-wide expansion. Other American magazines, catering to the same predominantly male demographic as *Wired*, have not fared as well. Indeed, *Wired*'s growth rate in the period between June 1996 and June 1997 was unusually healthy compared with other publications pitching to a similar audience. Computer magazines such as *PC World* and *PC/Computing* have experienced somewhat more modest gains during this period, while men's lifestyle publications such as *GQ* and *Rolling Stone* either barely maintained their market share or experienced some losses in the same period. The business magazine *Fortune* experienced a loss of 5.2 percent and *Business Week* made only minor progress with 2.4 percent total growth in paid subscriptions. It seems that the content of *Wired* resonates with the well-heeled segment of the male population more than these other periodicals. Although *Wired*'s circulation is only a small fraction of the rates enjoyed by these long-standing mainstream periodicals, the consistently strong growth of the magazine's circulation indicates that it has become a publishing force to be reckoned with.[13] In fact, *Wired* has even attempted to create an international mini-empire for itself. Since its inception in 1993 it has established the popular online version of itself, *HotWired*, launched a series of Hardwired books, provided its own political coverage in the form of *The Netizen*, started an online news service called *Wired News*, and published an editorial handbook called *Wired Style: Principles of English Usage in the Digital Age*. Wired Ventures has also spawned its own online search tool, Hotbot, and even attempted to tap into the television market with *The Netizen*, a short-lived newsmagazine show appropriately found on the newly created MSNBC — a joint project of Microsoft Corporation and NBC. How does *Wired* do it?

PHASE ONE: THE GENIUS OF *WIRED*'S MARKETING STRATEGY

According to *Wired*, the magazine creates an audience and niche for itself in the overlap created by *Scientific American* (a science magazine), *Forbes* (a business publication), *Premiere* (an entertainment periodical) and *Details* (a men's magazine).[14] In many ways, the promotional, highly commercialized character of *Wired* makes its content more similar to women's magazines such as *Glamour* or *Cosmopolitan* than to men's magazines. Though the technologically driven, masculine associations of *Wired* would seem to make such a comparison ridiculous, there are many significant similarities. Fashion magazines consistently privilege the artificial over the "real" and attempt to create images of the self that can only be achieved through consumption. In this way, fashion magazines create ideals of femininity that bear little relation to "natural" female characteristics and that demand intensive consumption of numerous products and an adherence to specific gendered codes of behaviour. *Cosmopolitan*'s "Cosmo girl" is an excellent example of this phenomenon. A similar dynamic occurs in *Wired* magazine — one becomes "wired" and can identify with the images of "hypermacho" masculinity found in *Wired* by consuming high-technology products and becoming a pioneer in the new digital frontier. In this way,

Wired, like fashion magazines, constructs and supports a particular, narrowly defined gender identity as a means of creating loyalty to the magazine and to the ideas it presents. It also uses numerous "advertorials" — a cross between a product advertisement and an editorial — that are very similar to the offerings presented by the beauty and fashion editors of fashion magazines. Advertorials are rarely critical and promote a wide range of products often without specifying a brand of choice. Persistent consumerism and a preoccupation with visual presentation are hallmarks of *Wired* and mainstream women's fashion magazines. Most importantly, both promote an ideology that is compatible with certain sexist assumptions and consumption expectations. In this sense, both have a stake in creating and maintaining unequal power relations of gender. While fashion magazines invite women to emulate their models and extol the virtues of beauty and deference to masculine authority, *Wired* uses complementary images of masculine power and technological supremacy. In doing so, *Wired* does the seemingly impossible: it transforms images of men formerly associated with computer geeks and uptight "suits" into ideals of hypermachismo. As we will see, *Wired* also reinforces a hypermodern version of the same traditional constructions of femininity that are found in women's fashion magazines.

This comparison provides important insights into how *Wired* fits into an existing hegemonic discourse of inequality and why women may "hate" the magazine. However, it fails to adequately recognize the significance of *Wired*'s content (which is dominated by the computer and telecommunications industry) and its unique design and ideological disposition. The ideological content of *Wired* goes beyond masculine technological consumer fetishism. To some degree, *Wired* seems to have created its own genre, as a hybrid of disparate elements found in other periodicals, including fashion magazines. This does not mean *Wired* is in any way inclusive and published for a general audience. Rather, it is designed for a particular, exclusive, high-end market as indicated by its price (US$4.95, $5.95 each in Canada) and by its large, colour-saturated, 9 ½ x 11 format. What it does mean is that *Wired* has targeted an audience in a particularly innovative and proficient way that has realigned male readers of other periodicals in its favour.

The genius of *Wired* and the secret of its success has been its ability to create a coherent, identifiable readership and deliver this captive audience to aggressive sophisticated advertisers. *Wired*'s readers are defined as members of the digital generation, the top 10 percent of corporate, government, media, education and technology elites who are perceived to be the "change leaders of our time" because they directly influence technology purchasing decisions.[15] However, the digital generation is also a fiction, a marketing tool created by the magazine itself. The concept serves as a kind of high-tech carrot in at least two ways: (1) it attracts new readers who want to learn about or who aspire to become part of the digital generation, and (2) it woos advertisers anxious to sell products to this presumably lucrative market. The construction of the digital generation as an elite class of the future and "the Ultimate Business Buyers" has created spiralling growth patterns. Yet *Wired* has not engineered its success all by itself. It has been the opportunistic beneficiary of the hypermodern climate outlined in Chapter 1 — a climate which repackages modernity's faith in scientific "progress" within the context of a capitalist system in overdrive. Though digital discourse is clearly not the independent creation of *Wired*, the magazine has carved out a niche for itself in the production and dissemination of a particularly dramatic and appealing form of this discourse. It has been able to identify the most sensational and engaging aspects of recent developments in new information and communications technology and reintroduce them to a male constituency that craves visions of technotopia and is committed to technological determinism. *Wired* is thus singularly promotional: it promotes itself, the digital generation it has constructed and the products it advertises. It also promotes the culture and ideology of a digital world in a self-referential, unabashedly commercial and future-oriented way. To understand *Wired* and the digital generation is to understand an increasingly powerful current in North American society.

The content that has attracted readers to *Wired* (and according to projections will continue to do so) is subject matter broadly related to developments in new information and communications technology. The dual developments of fibre-optic cable and digitization have expanded the potential of these technologies, and their expanded potential has increased the public's interest in the future of a technologically determined, information-rich society. *Wired* has capitalized on this heightened interest and has used existing currents in mainstream culture to motivate the public to "get wired," literally, by subscribing to the magazine, and figuratively, by becoming a part of the digital generation. *Wired* has sold itself by advertising its cutting-edge status and its ability to keep readers abreast of technological and social issues that it claims are not given adequate attention in the mainstream. Implicit in much of its advertising is a familiar threat: get online (or "wired") or get left behind. Many of its advertisements designed to lure new subscribers are less than subtle in this regard. For example, one ad features the caption "When the digital revolution rolls over you, you're either part of the steamroller or part of the road." Presumably, by subscribing to *Wired*, one can only avoid being run over by becoming a part of the "steamroller." These words are superimposed in colour on a black and white image of a woman clad in a retro-1960s lace top, underwear and horn-rimmed sunglasses and reclining on a lawn chair. A colour copy of *Wired* magazine has been placed in her hands. The combination of this image, reflecting a highly sexualized feminine image from a bygone era, and the threatening caption creates a series of possible associations. To *Wired*'s largely male demographic, it provides a perfect marriage of familiar, traditional gender images of the past with the hypermodern digital reality of the future. Just as new technology allows the advertisement's creators to alter the black and white photograph digitally so that it includes *Wired* magazine, so the digital revolution described promises the reader the power to remake reality in the future — to be "part of the steamroller" instead of "part of the road." Potential readers are thus invited into *Wired* as a challenge, as a dare to reclaim their power over the future while reasserting their masculine privilege.

Like its advertisements, *Wired* offers a particular kind of hypermodern content that ridicules the "old-fashioned" and holds out the possibilities of technotopia. Regular features consistently document late-breaking developments in the world of high-technology, interview the men (and very occasionally the women) behind these developments and explore possible scenarios for the future. Virtual reality is a common theme and discussions abound about how these lifelike, computer-generated artificial environments can offer the ultimate escape from a messy world of conflict. Other popular topics cover artificial intelligence, the latest in computer software and hardware, telecommunications policy and corporate ventures and mergers. Frequent contributions from leading scientists, technologists, technophilic social commentators and corporate leaders are prominently placed to prove that digital Nirvana is just around the corner. Discussion of future scenarios is so popular, in fact, that in January 1995 *Wired* published a special edition called *Scenarios*, which was devoted exclusively to exploring various constructions of the future as envisaged by key members of the digital generation.

This preoccupation with the future is one of the most significant elements of *Wired*. *Wired* does not simply report on business and computer industry developments, it also provides opinions on digital trends and generates enthusiasm about ongoing technological developments. *Wired* is selling the future and a big part of that future involves satisfying the consumption needs of computer and telecommunications industry elites. Approximately one-third of *Wired*'s content is advertising — either in the form of explicit high-quality advertisements or advertorials that feature text-driven "reviews" and product information.[16] Major repeat advertisers ensure the growth of *Wired* and provide the necessary funds to develop the magazine and increase circulation. Key advertising spaces have been continually filled by large corporate sponsors. Indeed, *Wired*'s 1996 Advertiser List reads like a who's who of the computer and information technology sector. Repeat advertisers include major computer and software companies (Microsoft, IBM, Apple Computer, NEC, Digital Equipment, Motorola, Compaq, America Online); computer game and CD-ROM corporations (Virgin Games, Nintendo); office equipment manufacturers (Canon, Hewlett-Packard); entertainment industry suppliers (Sony, Hitachi); high-end car manufacturers (Mercedes, Chrysler); and alcohol producers (Dewars, Remy Martin, Absolut). In view of *Wired*'s demographics, it is not surprising that it does not advertise conventional "women's products," like cosmetics, cleaning products, feminine hygiene items and so on. Although in many cases advertisers use the same ads in *Wired* as they do in other media, a striking harmony exists between the design and content of the ads and *Wired*'s editorial content. There is clearly a strong compatibility between *Wired* and its advertisers. While the ads provide detailed product information and create brand name association, the editorials of *Wired* help to construct the image of the digital generation that creates and uses such products. They also speculate on future developments and contribute to an overall message of a technologically determined future.

The close relationship between *Wired* and its advertisers is further indicated by the fact that in addition to the magazine offering its advertisers a captive and growing upper-class audience, the magazine provides its repeat advertisers with free original market research. Advertisers are offered access to *Wired*'s subscriber list in order to facilitate direct-marketing campaigns and are invited to collaborate with magazine staff to co-ordinate promotional opportunities and participation in special events. Online advertising opportunities have also been developed by the advertising staff of *HotWired*, *Wired*'s online counterpart, in order to maximize the impact of advertising with *Wired*. A list of Adlinks has also been included in more recent issues of the magazine. Indeed, *Wired*'s 1995 Advertising Kit explicitly acknowledges the magazine as a "marketing partner" for advertisers.[17] *Wired*'s loyalty to and support for the rise of the corporate digital information technology sector is abundantly evident.

PHASE TWO: MORE THAN JUST MONEY?

Since *Wired*'s "digital generation" is clearly a very attractive marketing tool for both advertisers and potential readers, it is not surprising that the magazine has grown from a small periodical embraced by a select group of San Franscisco technology enthusiasts to a virtual empire. But not everything *Wired* touches has turned to gold. Although the success of *Wired* magazine is truly remarkable, many of the subsequent spin offs of Wired Ventures have not enjoyed such good fortune. Wired Digital, the branch of Wired Ventures Inc. online, has reported huge financial losses. Despite the fact that the magazine's online equivalent, *HotWired*, has been critically acclaimed and is one of the busiest Web sites on the Internet, profits have remained elusive. Likewise, repeated attempts to take Wired Ventures public have failed to generate enough capital to sustain some of the company's many ambitious projects. For this reason, Wired Ventures has had to do some downsizing of its own, shutting down the operations of *Wired UK* after only a few issues, and laying off 20 percent of its Wired Digital staff. In July 1997, *Wired* editor and publisher Louis Rossetto announced he would resign as CEO of Wired Ventures after a search for his replacement had been completed. To many, this suggested that the company's financial woes were even worrying some of *Wired* key players — despite Rossetto's denials. As of the March 1998 issue, Rossetto became the first to hold the new position of editorial director, leaving his job at the pinnacle of the magazine's masthead to the new editor-in-chief, Katrina Heron. He also stepped down as Publisher of *Wired*, in favour of Dana Lyon. Yet the July 1998 issue of *Wired* still named Rossetto as CEO of Wired Ventures.

Despite this recent turbulence in staffing, the embarrassment of a TV debut that made few waves and lasted only four episodes and continued financial pressure, Wired Ventures is far from throwing in the towel. Indeed, the company is due to receive a large infusion of funds very soon. On May 8, 1998, Advance Magazine Publications' Condé Nast Publications unit (part of the S.I. Newhouse publishing empire) announced that it would soon take ownership of the magazine. This is good news for Wired Ventures, which has been relying solely on the magazine's success to support its struggling online enterprises. Capital from the sale of the magazine will reportedly be used to pay off the debts of Wired Digital, which has become Wired Ventures' fastest growing division.[18] Nevertheless, the willingness of Wired Ventures to operate at a deficit for so long suggests that there is more to the "wired" phenomenon than meets the eye. Could it be that making money is not the only thing that matters to the people at *Wired*?

Strangely, yes. Wired Ventures is not simply concerned with turning a profit and surviving in a new, competitive global economy. The *Wired* mandate appears to be far more expansive and concerned with influencing the direction of the new digital economy to reflect its libertarian (anti-government, pro-free market) values than it is with generating its own revenue. The financial success of Wired Digital may be less important than its role as a leader in the new digital frontier. The view of *Wired* as a magazine with loftier motivations gains some credibility when we consider its many friends in high places. According to Douglas Rushkoff's February 1997 article from the *Guardian Online*, the majority of the editors and experts on *Wired*'s masthead either are, or soon will be, associated with The Global Business Network. Kevin Kelly, John Perry Barlow and Nicholas Negroponte to name a few, have ties to this business think tank and consulting firm. Corporate members of this elite club include more than one hundred of the world's leading companies, 60 percent of which are predominantly large multinational corporations with headquarters in North America.[19] The Global Business Network is interested not simply in short-term profits but also in understanding and influencing future directions in the economy in the interests of business. The Network acts as a consultant for business in the present and as a consultant predicting and co-ordinating interests in the future. *Wired* appears to reflect a similarly long-term vision — something more akin to a religious mission than to a short-term business plan focused on the bottom line.

PHASE THREE: OH, WHAT A TANGLED WEB ...

Wired may find it has yet to face many of its most difficult challenges. While the magazine's ideological loyalties clearly remain, over the course of *Wired*'s short history its explosive success and growing readership have contributed to a change in its advertising profile. As the readership has grown, there has been an upsurge in product advertising not directly associated with the high-tech corporate world. Guess, Calvin Klein, the Gap and other increasingly diverse designer commodity producers have begun to infiltrate *Wired*'s advertising clientele, which was previously almost exclusively technology-related. As a result, the magazine has become increasingly mainstream in appearance — at least in terms of some of its advertising spreads. The dizzying success of *Wired*, then, has begun to taint the carefully designed, exclusive package that made *Wired* successful in the first place. At the same time, the libertarian, fervently pro-technology messages that have given the magazine its identity are becoming more and more common beyond its pages. With the growing success of *Wired* and digital discourse more generally, the cutting edge is becoming the mainstream. Its success has, ironically, created potential problems for the periodical.

Less than 40 percent of *Wired*'s subscribers actually hold top management positions and are thus "genuine" members of *Wired*'s exclusive digital generation. A greater number of readers and subscribers do not belong to this group. Though these individuals may identify with the digital generation and see themselves as "wired," the elite class *Wired* purports to cater to is a somewhat deceptive category. That is, the digital generation is both a real demographic category and a fabricated identity designed to draw people into the magazine. If *Wired*'s circulation continues to grow, the proportion of readers that comprise the most exclusive, upper echelons of the digital generation will decrease as the proportion of readers aspiring to join this class, or at least identify with it, increases. Hence the problem *Wired* faces is in its success: can it maintain the exclusive digital status it covets if its readership becomes too diffuse? Will *Wired* continue to be *Wired* or will its numerous recent partnerships force the magazine to water down its content? Will *Wired* be destroyed by its own success? Some watering down is probably inevitable in the life of a so-called cutting edge periodical like *Wired*. This is especially the case in view of *Wired*'s recent purchase by the publisher of such mainstream magazines as *GQ* and *Vanity Fair*. However, it is important to recognize that the magazine's success has also allowed it to influence hegemonic discourse more forcefully — *Wired* may simply appear less intense because mainstream discourse is becoming more and more wired itself.

It remains to be seen whether *Wired*'s readership will become less exclusively male. Some movement in this direction has occurred. According to *Wired*'s subscriber profiles, female readership grew from 12.1 percent in 1995 to 18.9 percent in 1997. Little is known about how female readers view the magazine or whether they identify with *Wired*'s message in the same way as its predominantly male readership. Borsook's informal survey suggests they do not. Making *Wired* more "woman friendly" is not simply a matter of ensuring that some women are involved with the production of the magazine. Although the vast majority of *Wired* content is written by men and clearly appeals to a largely male audience, women have participated in the creation of *Wired* since its first issue — even the president of *Wired* is a woman (Jane Metcalfe). Regardless of the sex of the magazine's producers, the exclusive, masculine nature of *Wired* discourse remains undeniable. According to Borsook, approximately 15 percent of *Wired*'s authors are women, but the majority of articles by women that are accepted for publication focus on sex and dating or are features about other exceptional wired women.[20] The issue is not that there is a lack of women technical writers or a shortage of women submitting articles to the magazine, it is an editorial one. Borsook has revealed that attempts to integrate gender analysis or woman-centred issues into some of her submissions to *Wired* met with consistent opposition. She argues that the lack of women in senior editorial positions is the key problem. This is doubtful. The replacement of Louis Rossetto with Katrina Heron as editor-in-chief has not translated into significant changes in the magazine's content. Gender issues remain ghettoized and traditional gender stereotypes have been maintained. This suggests that specific, if unspoken, editorial criteria ensure that *Wired* maintains its unique, masculine image, regardless of who is at the helm.

It seems that *Wired*'s exclusive status is safe with respect to gender. Still, the question remains: Why do the few women who do read *Wired* do so when so many others, according to Borsook and the readership's demographics, are put off by *Wired*? Is there something exceptional about these women? Borsook admits that writing for the magazine did bring with it a certain cachet: a sense that one belonged to a very privileged group.[21] Women readers may experience a similar feeling, based in part on the magazine's exclusive masculine association. Other possible explanations can be found by looking at the demographic information provided in *Wired*'s Advertising Kit, which includes two surveys: one focuses on all readers and one focuses exclusively on male readers. The logic of producing a separate male-only list is not clear, especially in view of the fact that *Wired*'s readership is over 80 percent male anyway. The differences between these two lists are negligible: the average age remains just over thirty-nine years, and employment remains heavily clustered in the professional/managerial sector, followed by the self-employed and top management sectors. Not unexpectedly, however, both employment and household income averages are slightly lower in the survey of all adults than in the survey of male readership alone. In addition, education levels are slightly lower in the unisex list, suggesting that the few women who do read *Wired* are likely to possess unusually high levels of education. The main difference between these statistics is found in the category of postgraduate degrees. The adult reader profile indicates that 31.9 percent of male and female readers have postgraduate degrees, while only 23.5 percent of male readers have attained such a level. There are many possible explanations for why the minority of women who are attracted to *Wired* are also highly educated. One of the most likely is that women with extensive academic experience are more likely to be familiar with the technologies with which *Wired* is the most concerned, and therefore may not be so likely to feel as excluded by technological discourse in general. These women remain in the minority, however. Women's historical exclusion from technology, it would seem, endures.

A MACHINE THAT GENERATES GENDER

In view of the demographics of its audience, it is not surprising that the *Wired* machine concentrates its energy on generating images that assert a rigid form of powerful, digital masculinity. *Wired* visually reflects images of its readership or fantasies of what its readership would like to be. Yet the dominance of this male-centred imagery, which is explored in Chapter 6, does not mean that *Wired* is uninterested in femininity. In the dualistic world of patriarchal, normatively heterosexual culture, femininity remains of great interest to men as readers. It is also important to understand that the definition of masculinity is made in part in opposition to femininity. Men are defined as men because they are *not* women. This is nothing new. The history of western culture, from ancient Greek myths to the popular *Sports Illustrated* Swimsuit Edition, reveals the need of dominant masculine-oriented culture to define, control and manipulate images of women.[22] In this way, images of femininity and otherness become a foil to the dominant, idealized identity and serve to enhance its status. This is evident in the case of *Sports Illustrated*'s Swimsuit Edition, which uses stereotypical constructions of femininity to enhance the popularity and masculine associations of the subject — in this case, the world of competitive sports. In *Wired*'s case, images of women and minorities perform a similar function for the world of new digital technologies. Digital discourse has recognized and reproduced an important "truth" of western culture: one of the best ways to elevate something's status is to associate it with white masculinity. The association of digital technologies with white masculinity, then, enhances their status in western culture. It also helps to create a separate, exclusive culture of digital technologies. As a result, understanding how *Wired*'s discourse constructs difference will help us to understand exactly what interests are being promoted by digital discourse and how this unique wired world is created. Only then will we be ready to turn our attention to how the dominant masculine ideal of *Wired* — the hypermacho man — is constructed.[23]

Chapter Four

OUT OF THIS WORLD: EXCLUDING, RECONSTRUCTING AND ELIMINATING DIFFERENCE

> The twentieth-century ends with the growth of cyber-authoritarianism, a stridently pro-technotopia movement, particularly in the mass media, typified by an obsession to the point of hysteria with emergent technologies, and with a consistent and very deliberate attempt to shut down, silence, and exclude any perspectives critical of technotopia. Not a wired culture, but a virtual culture that is wired shut ...
>
> — Arthur Kroker, Michael A. Weinstein, *Data Trash: The Theory of the Virtual Class*

> A firm must be totally dedicated to its core values. This requires a "cult-like" atmosphere in which employees are "indoctrinated" into a "corporate ideology" with misfits being "expunged like an antibody."
>
> — James C. Collins and Jerry I. Porras, as quoted by Phil Agree in "What Makes a Vision?," *Wired*, March 1995

TO ENTER THE *WIRED* WORLD is to intrude upon the elite digital generation's presumed hypermodern future. It is to peer into the world beyond the current period of rapid technological change and glimpse the future envisioned by the "change leaders of our time." It is a world that is foreign to many of us, a world constructed in grand artificiality and populated by a tightly controlled collection of digitally enhanced beings. One of the important ways that *Wired* attracts its readership and constructs its particular form of digital ideology is through the representation of difference as social categories that can be either excluded from *Wired* discourse, constructed in nonthreatening ways or, ultimately, eliminated altogether. Through these multiple, contradictory techniques, *Wired* discourse constructs an artificial world that continually sends the same basic message: the digital age is the domain of a masculine digital elite. How does *Wired* do this? First, it excludes those it considers Other — women, minorities, the poor, the technologically illiterate — to create an exclusive, high-status masculine world. The magazine then populates this artificial world with images that reinforce existing hierarchies and support its highly individualistic ideology. Finally, by invoking images of technotopia, *Wired*'s discourse puts forward a vision that suggests social difference may disappear altogether in the virtual world of the future. In this chapter, we will begin to explore the artificial world created by *Wired* magazine and examine how it serves to reinforce sexist and racist assumptions through each of these techniques. We will see that although *Wired* magazine espouses the rhetoric of social transformation, it actually perpetuates long-standing inequalities, as do other forms of digital discourse.

FLASH AND DAZZLE: CREATING THE EXCLUSIVE WIRED WORLD

In view of western culture's shift from a text-based society to an increasingly visual one, it is not surprising to find that in order to command attention in our media-saturated culture one must have an original and intense visual design. This type of design has made *Wired* a veritable showcase of print media technical savvy that caters to a culture waiting to be dazzled by the products of cutting-edge technology. It is important to understand how the magazine uses visual techniques because they are crucial to how *Wired* creates a synthetic world on its pages. This world visually excludes many from the discourse of digital culture before a single word of text is read. Understanding how *Wired* does this also helps us to understand why the magazine appeals to a particular male demographic and excludes others. How is this distinct, futuristic world created and what is its purpose?

Out of this World: Excluding, Reconstructing and Eliminating Difference

Wired magazine undoubtedly has an innovative and unusual look. In fact, its design, colour, layout, graphics and typography are essential to understanding the magazine's content and appeal. These elements allow *Wired* to achieve three main goals: (1) to command the attention of a visual culture, (2) to create the visual representations necessary to construct the digital generation as an exclusive social group and (3) to promote and reflect an emerging digital ideology that includes a particular set of gender and race constructions. While this chapter focuses entirely on *Wired*'s unique design sensibility, it is important to remember that *Wired* is only one expression of a larger digital culture. Similar design elements are found elsewhere in North American popular culture. One example is the newly revamped version of *The Globe and Mail Report on Business Magazine.* Although this magazine does not share all of *Wired*'s digital ideology, its design emulates *Wired*'s trademark style. A similar design sensibility is also found in the many current television ads that barrage viewers with rapid, complex and intensely coloured images.

Wired style can afford to be more intense than either of these examples because it is designed to appeal to a more select and narrow audience. The magazine structures its content to appeal to the technical elite; a group for which comfort with all things digital is a defining characteristic. Reminiscent of an online Web site or Windows computer interface, *Wired* offers its readers short editorial pieces, many of which are presented graphically in the form of lists, tables and "infonuggets." The layout, which varies from page to page and integrates diagrams, photographs and text, is familiar to users of multimedia products and graphic user interfaces. These digital products present viewers with a number of icons that represent the subject matter available when they are selected. The *Wired* layout functions in a similar way by facilitating the display of many related pieces of information on one "screen" (or page), while continually stimulating the reader visually and encouraging short-term engagement with specific pieces of text. The reader is bombarded with a visual display on each page and a series of text-based "windows," which she or he may either read, glance over or skip altogether by turning the page to a new and different screen. This type of design seems to incorporate elements from various magazine genres, including tabloids, newspapers, comic books, as well as "traditional" consumer magazines, with the Windows-based technology of the computer age. The result is a hypertextual experience that encourages brief, superficial engagement and reflects a hypermodern world characterized by speed. The disjointed structure creates a unique reading experience that continually focuses and refocuses the reader's attention, engaging the reader intuitively and visually as much as textually. While this format does not demand a long attention span or critical reflection, it does demand a particular sensibility — one that is comfortable with a hyperactive style and with the world of digital technology.

The intense, hypermodern design of *Wired* is reinforced by digitally altered photographic images that blur the line between the real and the artificial. Technologically mediated art dominates the artificial *Wired* world and frequently flirts with the "hyper-real" — the technological exaggeration of real-life phenomena almost to the point of rendering them unrecognizable. Intense colour is central to the creation of this hyper-real design. Both the cover and the internal content of *Wired* are filled with rich, stark, saturated, profoundly "unreal" colour. Extreme combinations featuring either hot colours (reds and oranges) or cold colours (blues and greens) are often used. *Wired*'s colour palate is predominantly composed of a variety of neon shades. Tones traditionally associated with the natural world, such as earth brown and forest green, are almost entirely absent. The viewer is confronted with grass that is greener than green and skies that are filled with perfectly formed, symmetrical clouds. It is also unusual to see a flat colour background in the magazine; texture or graduated colour is common. In addition, colour is not limited to the background and graphics, but is frequently used to enhance portions of text and unify sections thematically. This dramatic use of colour and numerous fonts makes this highly stylized magazine immediately recognizable. It also makes the reader associate *Wired* magazine with a futuristic, computer-mediated reality. This invokes many of the long-standing myths of modernity that continue to hold sway within our hypermodern culture. From telecommunications ads that show parents sending pictures of their developing fetus to relatives via the Internet, to depictions of messages being faxed to Earth from three thousand feet, we are invited to glimpse the digital world everyday. As a result, the world of *Wired* is both vaguely familiar and strangely foreign to those of us who do not fit the demographics of its primary readership. The promise of technological progress and technotopia are suggested by depictions of a manmade, synthetic reality that conform to the latest notions of high-speed technoculture. In this way, the graphic designers of *Wired* create a world that is separate from the "real" world beyond its pages: an exclusive, masculine techno-world that cannot help but be noticed.

Wired also includes a large number of digitally enhanced, altered and generated photographic images. In fact, it is difficult to find a photo that has not been altered, at least to some extent. Photographic images are most frequently blended with others, combined with computer-generated graphics, or enhanced through the intensification of natural colour or the addition of artificial backgrounds or artificial effects. The result is a magazine that depicts the world, even the "natural" world, in a completely artificial, fantastic way. While certainly these techniques have been used to a lesser degree in advertising for some time and are familiar to consumers of air-brushed, soft-focus fashion photographs, the dominance of all content by an overtly technologically mediated style is remarkable. In using this visual technique, *Wired* creates a new futuristic world and promotes a design sensibility of virtual superficiality and hyper-reality. As a result, we see bodies infused with neon colour, and cartoons, machines and people interacting on a futuristic, artificial background that evokes cyberspace. This image of the future is completely computer-mediated and is consistent with a digital ideology that espouses the virtues of technotopia.

Wired's photographic strategy differs dramatically from magazines such as *Life* or *Time*. Those publications aim to present images that reproduce reality so that the viewer "forgets" that the photograph is separate from the reality; signified and signifier are intentionally blurred. However, *Wired* is not designed to reflect existing realities, but to influence future directions. This intention is indicated by *Wired*'s explicit desire to get subjective writing from its authors while serving an elite audience and, in the words of the former *Wired* editor Constance Hale, continuing to "reflect our fascination with science and technology."[1] In fact, *Wired* creates an artificial *Wired* world on which the "real" world is to be modelled. Its photographic strategy is, therefore, the opposite of traditional news periodicals. When we view an image in *Wired* we must decode it in this context. We must not ask ourselves "What reality is this image trying to capture?" but "How are *Wired*'s views of the future being reflected by this image?" Or, more pointedly, "What does this image tell us about the future world imagined by the digital generation?" Just as the majority of *Wired* discourse is anticipatory and future-oriented, so are the graphics it presents.

Not surprisingly, the computer keyboard and the machine to which it is attached dominate many of *Wired*'s graphics. The future is, afterall, digital. Images of computers, printers, faxes, modems, software and miscellaneous technological paraphernalia abound. Frequently, the device in question is presented dramatically, complete with spotlight, attractive background and intense colour enhancement. These devices generally assume centre stage in the image. In addition, the fascination with technology that imbues *Wired* is reinforced by using clip art and graphic borders that incorporate computer circuitry, models of bits and complex cable systems to decorate the articles. Whether or not such images directly relate to the text they surround, they suggest the existence of a computer aesthetic, which appeals to *Wired*'s presumed audience — the digital generation. This style is well known and highly researched by computer designers. The dominant image is of a sleek, symmetrical, box-like product, typically white and accentuated by a shiny black screen or texturized surface. The simple outer design suggests inner complexity, which is often graphically represented by the graphic border or clip art. Small size is usually valorized, so the object is often seen in relation to a male hand. This small size is also juxtaposed with descriptions of the product's virtually limitless potential. It is not unusual for the actual product to be absent from the ads altogether: the focus is on the enormous capability of the product, as indicated by a dramatic graphic of cyberspace or digital art. Regardless of the particulars of the presentation, the message is all too frequently the same: technology is both beautiful and powerful and the key to the digital future.

What is the appeal of such a format? Aside from the minimal demands it makes of its readers' critical abilities, *Wired*'s hypermodern, technologically mediated format also visually separates the magazine from other publications. It signals that *Wired* is something different; that its content is designed for the uniquely contemporary, technologically savvy, cutting-edge members of a new digital culture. *Wired* endeavours to make its specialized digital knowledge "hip," exclusive and high status. As a result, the magazine's features have catchy, informal titles like "Street Cred" and "Net Surf" that are unique to *Wired* and familiar only to its digital generation readership. "Rants and Raves" is a regular column featuring letters to the editor. "ElectricWord" contains short "bulletins from the front line of the digital revolution," and includes the popular "Tired / *Wired* List (which categorizes various elements of popular technoculture), "Jargon Watch" (which provides definitions for the latest digital slang) and the "*Wired* Top Ten" (which rates aspects of high-technology culture). The column "Fetish" introduces new products, complete with prices and phone numbers for ordering purposes, and provides what *Wired* accurately calls "technolust" every month. The "Hype List" and "Reality Check" ostensibly distinguish between actual and artificial technological trends, while "Deductible Junkets" updates the digital generation on upcoming technology-related events and conferences. "Raw Data" uses a visual format to illustrate such data as investment patterns, directions of scientific research and success levels of various companies. "Follow the Money" explores technology related investment issues. "Geek Page" provides a short, highly technical discussion of new technological developments that assumes a high level of scientific and computer knowledge. More recent sections like "Just Outa Beta" previews potential new products, and "Updata" updates previous stories. Finally, "Electrosphere" covers developments in the electronic, online world of the Internet, and "The Netizen" provides coverage and analysis of U.S. federal politics for the digital generation. *Wired*'s most overtly ideological pieces are found in the longer feature articles and interviews that vary from issue to issue; in regular features such as "Idées Fortes" (a collection of short, often controversial, pieces); in the high-profile "Cyber Rights Now" or "The Netizen"; or in the senior columnist Nicholas Negroponte's editorial last word. Together, these pieces allow *Wired* contributors and editors to flex their ideological muscle; to ruminate about the relationship between man, machine and god;[2] to speculate about the liberating effect of the digitization of public libraries;[3] to provide a regressive perspective on immigration policy;[4] or to discuss information technology's battle to "free" China.[5] As we shall see, these articles reveal a consistent digital ideology that ignores or denies the significance of issues of inequality that pervade western industrial society.

Meanwhile, *Wired*'s graphic style marries elements from the past with its technologically produced future, visually excluding threatening difference from its hypermodern world. One of the most significant ways that this is accomplished is by blending different styles and images from various eras in commercial art history. By mixing up genres from each of the decades between the 1950s and the 1990s, *Wired* creates an eclectic look and feel. Psychedelic imagery associated with the 1960s is juxtaposed with computer-generated science-fiction art of the 1990s; feminine caricatures from the 1950s are repeatedly invoked. The blending of these diverse styles, always with the mediation of high-tech production methods, is quintessentially *Wired*. While it sometimes appears whimsical and playful, this complex combination of styles is far from wanton. Rather, it represents an attempt to visually unify distinctive time periods and dehistoricize the present/future world of *Wired*. In other words, *Wired*'s world begins in the present and proceeds to the future: the past is relevant only as a source of particular images that convey messages that digital discourse seeks to transport into the future. Because the ideology underlying digital discourse is future-oriented, a determined effort must be made to reinforce the notion of progress and suppress traditionalist impulses that resist technological change. By integrating specific elements of the past into a hypermodern present/future, *Wired*'s design provides its viewers with an important visual message about the past and the future. The juxtaposition of futuristic imagery with technologically mediated graphics that glorify select images from the past allows *Wired* to reassure its readers that while the future will certainly be different from the present, it will not be beyond the control of the elite digital generation. Thus, as digital discourse keeps our eyes fixed on postmodern imagery and the rapid technological change of the present/future, it re-creates traditional sexist and racist stereotypes. Between the flash and the dazzle, the power relations of the future are being reinscribed in all too familiar ways.

MOTHERS, CYBER–FEMME FATALES AND GANG MEMBERS: THE RECONSTRUCTION OF DIFFERENCE IN *WIRED*

The way that *Wired* discursively constructs the Other is of great importance to feminist discourse analysis. The constructions of women and minorities that are found form a separate discursive stream in *Wired* and are relegated to subordinate status. However, the images and text that represent women and minority groups do play an important role in the creation of *Wired* culture. They become more significant as a result of their rarity. Throughout the advertising, editorials and advertorials of *Wired*, there are many familiar stereotypic visual constructions of race and gender. Present on the fringes is the female sex object; the non-threatening female secretary or lower management type; the silly, carefree suburban housewife; the mother; the schoolgirl. There are also occasional images of the macho, musclebound Black male; the exotic African; the exceptional middle-class, business-suited Black man. Before examining some of these images and how they perpetuate specific power relations of race and gender, it is important to look at how *Wired* itself has raised the issue of gender roles. How it does so serves as a backdrop to *Wired*'s presentation of gendered images.

Unfortunately, the result is frequently the reassertion of conventional sexist views. One of the most vivid examples of such rhetoric is found in *Wired*'s November 1995 issue. In the high-profile four-page illustration at the beginning of the magazine, *Wired* highlights the following quotation from a feature article:

> Two styles of people: guys and gals. Females, what? They nurture. Men, what? They squirt and move on. So, business start-ups — same thing. The entrepreneurs who run businesses? They're like women. Caretakers. Venture capitalists, though? Gigolos. Roosters. Seed capital. Get it?[6]

The accompanying graphics of the first two-page spread reveal computer-generated images of middle-aged white men, either breast-feeding child-size robots or tending a digital garden. (See Figure 2.) The second page reveals similar computer-generated roosters manipulating robots in an artificial field of pipes and circuitry. Although the article quoted in this spread has little to do with gender — it explores the lives of fictional Silicon Valley venture capitalists — it is this harsh, stereotypic quotation that is highlighted. While certainly not the only instance of overtly sexist views, the starkness of this quotation is nevertheless significant. Intended to get the attention of readers and to "sell" the satiric story, the offensive remark combines sexist attitudes and a free-market, mechanistic perspective with the simplistic, aggressive style of much of *Wired* discourse. The effect is to re-create solidified gender relations of the past in a cybernetic age of the future. Interestingly, *Wired Style* editor Constance Hale later commented that some readers never recognized this feature article as "well, satire about Silicon Valley."[7] Hale does not explore why. She simply notes that it is not *Wired*'s practice to label fiction as fiction. However, the fact that the article explicitly articulates assumptions expressed in less dramatic (or satiric) fashion throughout the periodical makes such reader interpretations understandable.

Wired's sexism is also evident in the way it portrays the few women that it chooses to feature. Since 1993 only two real women, Laurie Anderson and Sherry Turkle, have appeared on the cover of *Wired* magazine. In both cases, *Wired* emphasizes their gender and uses various techniques to remind readers of their exceptional status in a male *Wired* world. The political consequences of their gender are avoided: they are simply unusual women who are able to function adequately in a digital world. Dubbed "America's multimediatrix" and carefully distanced from feminism, the musician Laurie Anderson is described as a self-proclaimed "techno-ice-queen observer." She is a highly supportive *observer* of information technology ("the campfire around which we tell our stories") and does not challenge its power. She even refers to the Gulf War as "absolutely the biggest multimedia spectacle in the last few years," ignoring the political issues surrounding the war and the human tragedies of the event. Like the true "techno-ice-queen observer" she proclaims herself to be, she reveals no human connection at all to the violence of the war. The subject of her work, meanwhile, is described as "what it means to be an American adult today." Class, gender and race issues are erased. In this way, Anderson is simultaneously sexualized as a multimeditrix, and desexualized as simply another supporter of information technology. In a *Wired* world, gender only matters if women are present (and they usually aren't). If women are present, this phenomenon must be accounted for in a way that preserves the integrity of the *Wired* world as a masculine environment.

A similar dynamic occurs with Sherry Turkle, who appears on the cover of the April 1996 issue. Like Anderson, she is portrayed in true

Figure 2: In the high-tech virtual world of the future, traditional gender stereotypes are asserted in stark hypermodern form. (This graphic accompanied Po Bronson's article, "The Relentless Pursuit of Connection," *Wired*, November 1995, 11–12.)

Wired form, through a lens of techno-static. Neither women look at the camera head-on, but are shown with heads tilted. They are certainly not the challenging male figures found on *Wired* covers. The images of Turkle contained inside the magazine are cast in soft blues and pinks and are anything but aggressive: the subject smiles benevolently at the camera, with gold earrings glinting. The article about Turkle in this issue is also stripped of its political relevance. The gender analysis that forms an important part of Turkle's work is ignored. Instead, the article, entitled "Sex, Lies and Avatars," subverts gender issues in favour of a discussion of cybersex. Like Anderson, Turkle is sexualized, with innuendo that suggests she "knows a lot about" cybersex, and desexualized as a scientist and postmodern theorist who studies genderless online behaviour in a nonthreatening, apolitical way. Despite the fact that most of the article focuses on Turkle's research and postmodern theory, the four-page spread at the beginning of the magazine quotes Turkle as follows:

> Is online sex like having an affair? Is it my business because I'm married to you?
> Or is it like you're reading pornography and it's none of my business?

> In our new questions about authenticity we see the beginnings of a cultural conversation that's going to take 50 years.[8]

Apparently, *Wired* readers are more likely to find a woman discussing cybersex more fascinating (and more believable?) than her sociopolitical research.

Wired uses similar techniques when describing the other exceptional women it occasionally features. For example, one article describes the remarkable skill of Christine Downton, a financial analyst whose knowledge of the bond markets has recently been uploaded into a computer system. Rather than portraying this woman as a stellar *Wired* individual, her gender immediately subsumes her achievements. The opening sentence reads, "A lot of men will tell a woman it's her mind they're after. But in the case of Christine Downton … it was true."[9] Meanwhile, the technology theorist Donna Haraway is presented to readers in the least threatening way possible: "soft-spoken, fiftyish, with an infectious laugh and a house full of dogs and cats … like a favorite aunt."[10] The rest of the article presents Haraway's views in a positive light, as pro-technology "cyberfeminism" that does not challenge digital culture.

This more subtle gender stereotyping is complemented by the conventional images of women that appear in *Wired*'s advertisements, once again proving that "sex sells." Frequently, only portions of the female body are evident: a torso behind a computer screen, a heavily made-up eye peering from a colour monitor. The July 1998 issue of *Wired* features a Lycos advertisement that shows a magnified image of a microchip alongside a well-endowed woman's chest in a bikini top. The two are separated only by a mouse cord. The text below reads, "Lycos will find exactly what you want. And then some."[11] Male heterosexual desire and technology are clearly linked in the ad. As is common in car advertising, computers are also directly compared to women: an IBM computer prototype is "as idealized as a fashion model, as thin and sensuous as Kate Moss."[12] Fragmenting women's bodies in order to emphasize their objectification and decorative status, and associating products with female sexuality to enhance their appeal, are typical advertising techniques. These techniques also perpetuate the feminization of technological artifacts as objects that are malleable in the hands of men. Yet the status of the computer as a gendered object remains somewhat ambiguous. As Springer notes, although "mechanical objects have been imbued with male or female sexual characteristics for centuries," the computer has been described both in terms of masculine characteristics (power and speed) and feminine characteristics (small size, internal, hidden workings).[13]

Despite this ambiguity, images of the "mechanical bride" (a term first used by Marshall McLuhan in the 1960s) emerge on the pages of *Wired.* These fragmented and mechanized images of women are an important element of what McLuhan called "the folklore of industrial man." They fulfill a long-standing cultural yearning to create the perfect artificial woman. Mechanical brides are important as a means either to expand sexual possibility through the use of technology or to sexually possess technology itself. In either case, McLuhan sees the "interfusion" of sex and technology as "one of the most peculiar features of our world."[14] In view of the fact that *Wired* has named McLuhan its patron saint, it is not surprising to find mechanical brides in the magazine. These images encapsulate many of *Wired*'s stereotypic assumptions and preoccupations. One such example appears on a recent cover of the magazine, which features a synthetic image of a cybernetic woman wearing virtual reality headgear and blue lipstick; the woman stares blankly at the reader.[15] This futuristic, cosmetically beautiful woman looks at the viewer across an angle through layers of multicoloured eyeshadow. The accompanying article, a special report on "Hollywood 2.0," explores the growing role of new technologies in the film industry. It discusses such things as "post-human" talent agencies and renting movies on the Web in an entertainment world "where computer-generated actors are competing with flesh and blood."[16] There is no direct relationship between the cover image of a female and the internal subject matter. Rather, the feminine image reinvents the mechanical bride for the computerized entertainment industry. Another short article depicts another such synthetic woman, a completely "(Im)material Girl," described as "the first computer-generated pop star."[17] The creation of Japanese computer engineers, the virtual woman can apparently entertain in any language and potentially replace thousands of real female entertainers. The ultimate virtual woman, this "(im)material girl" can be manipulated to replicate the fantasies of her creator. The fact that the images of the cyborg creations are predominantly women reflects an interest in using new technologies to re-create the feminine within a defined set of parameters. If the images presented in *Wired* are any indication, these parameters closely resemble the feminine ideals of women's fashion magazines.

Wired's advertisements also construct the feminine image in specific ways. Because technology products are traditionally confined to the male domain, advertising must clearly distinguish between the user of the technology and the woman shown in the ad. While there are exceptions to this general characteristic (for example, a woman seated at a computer), even these often depict women in a subordinate secretarial role or merely an ornamental one. A common way in which high-tech ads involve women is through the invocation of the Greek goddess or celestial muse. Coupled with an image of the product hurtling through outer space or hovering powerfully in the heavens, a bald ethereal female figure is shown, for example, in a Sega Saturn ad, dressed in skimpy, gauzy draped fabric, holding out "the ultimate game system" against a background of turbulent, cloud-filled sky. This imagery arouses mystery, excitement and sexual desire, but the product remains firmly within the masculine sphere: a gift to men from the heavens.[18] Such images are far from novel and can be found in advertisements of many products in mainstream culture. It is also interesting that women rarely appear in groups in either *Wired*'s ads or editorials. According to Mary White Stewart, this omission is common in periods of insecure gender roles in which there is not a strong feminist movement and there is a high level of female objectification.[19] The current backlash culture demonstrates these characteristics. It appears that images of women in groups represent a threat.

Wired's images of women are not all simply reproductions of advertising techniques found elsewhere. Afterall, innovation is purported to be one of the hallmarks of digital discourse. While many *Wired* images of femininity do differ from those found in mainstream culture, they still continue to reassert familiar gender stereotypes. Most echo the virgin/whore dualism found in much of western literature. The virgin construction usually involves the reactivation of a stylized image of 1950s and 1960s American pop culture. Such images portray an idealized, very feminine, highly sexualized, marriageable (or married) woman. She is sweet and all-American, she has an hourglass figure, she is predominantly white and blond, she is often clad in styles of the 1950s or mainstream 1960s, she is very social. She is your mother, the girl next door and the idealized goddess of femininity. She personifies a glorified past when gender roles were understood[20] and adhered to, but exists in neon hypermodernity, giggling as she hands out advice about cybersex and social life, or selflessly serving male needs.

These "cyber-Barbies" are often textually constructed in traditional roles that are simply jettisoned into the present/future of the digital age. Articles such as "Coco's Channel," "The Mother of Multimedia" and "Cybrarian," for example, describe women in their traditional roles as teachers, mothers and librarians, while indicating how technology has and will positively transform their occupations.[21] (Such transformations are very limited, however, and do not upset the low status of these occupations.) Traditional "family values" are ascribed to these women textually. They are said to be concerned with the welfare of children, with education, with helping the researcher. Thus the subject of "Coco's Channel" sees herself "making the world a safer place for nerds," while "The Mother of Multimedia" is "the soothing parent who inspires her children to stretch." Computer hacker Carolyn Meinel is likewise described as a "concerned mother" who wishes to create a healthy environment in cyberspace for her children. The fact that these women are attractive and exude an "overwhelming Mom-ness" is also emphasized in the articles, which highlight their exceptionality in digital culture as the result of their gender. Advertisements such as IBM's OS/2 Warp promotion reinforce these roles by showing the only woman in the ad using IBM technology not for business purposes but to "warp photos of the kids to Grandma."[22] In a similar way, the Allison of *HotWired*'s now defunct "Ask Allison" column and the Diva of *Wired*'s semiregular "Netiquette" column re-create familiar gender roles in cyberspace. By providing flippant and lighthearted social advice to Net surfers, these women enter the digital age with traditional gender roles in tact. Even Miss Manners gets into the act in an interview on "netiquette" with the *Wired* writer Kevin Kelly. While Kelly is clearly not impressed with Miss Manners's traditional rules of conduct that favour "real life" interaction over online relations, her status as an expert on "excruciatingly good behaviour" is unchallenged.[23] Although these women provide some of the few female voices in *Wired*, their voices are clearly dominated by conventional views of women's roles in the larger society. They are another version of the common patriarchal virgin–mother: nonthreatening, subservient, domestic, beautiful and nurturing.

The images of a young woman from the 1950s or 1960s in advertisements for high-tech equipment associated with the future perform a similar function. Such ads depict cyber-Barbies in stereotypic poses, staring blankly out of the page, absently pushing a computer key,[24] or acting the part of a frightened 1950s horror movie victim — plagued by inadequate memory.[25] They are particularly significant because they not only fuse traditional images of women with technology but also meld a distant past with the future in an almost seamless manner. Like the image of the woman in the "steamroller" ad for *Wired* subscriptions (discussed in Chapter 3), these images seem to suggest that the gender roles of pre–second-wave feminism are compatible with the cybernetic culture of the future. They also suggest a return to a time when gender roles were ostensibly stable and unchallenged, thus providing a degree of comfort to an uncertain future.

The second uniquely digital construction of femininity is *Wired*'s version of the aggressive cyber–femme fatale — the whore to cyber-Barbie's virgin. She is the highly sexualized bad girl of cyberspace. Often clad in studded leather bustiers and chains or metal breastplates, she is the voluptuous cybersex fantasy, aggressively surfing the Net and appearing in a variety of computer war games. She is simultaneously presented as predator and prey, secretly waiting to be overpowered by a skilled cyber-warrior. Such figures are sometimes appropriations of women's images from the 1990s "riot grrls" movement. This movement seeks to celebrate girls' self-confidence and aggression. In *Wired*'s version, however, any original feminist meaning that may exist in "grrl" images is lost. The cyber–femme fatale, like the cyber-Barbie, allows *Wired* discourse to redeploy traditional gender stereotypes in a hypermodern, uniquely digital form. (See Figure 3.)

The cyber–femme fatale is depicted through advertising and articles as an aggressive sexual or violent predator who is waiting to be tamed. Computer game ads are among the most forceful purveyors of this construction; they frequently include images of scantily leather-clad women warriors, awaiting violent sexual predation.[26] Squaresoft's advertisement for "Final Fantasy VII for Windows 95," for example, features a cartoon-like image of an extremely well-endowed woman in a cropped top and holster with the caption, "You want a piece of me program boy?"[27] PlayIncorporated uses a similar technique, portraying its "technovangelist" Kiki Stockhammer, firing a futuristic weapon while wearing knee-high black buckled boots, fishnet stockings and blue metallic shorts. The caption above reads "Blaster Space Chick."[28] Short articles reflect similar themes, suggesting that cyberspace is filled with sexually promiscuous and available women, aggressively surfing the Net in search of hypermacho men. Articles such as "Send Dirty to Me," "Bianca's Smut Shack," "Babes on the Net," "Cyberdykes R Us" and "Quake Girls," for example, deploy traditional myths about women as highly sexualized objects, who forcefully seek male attention and crave sexual domination.[29] They are perfectly at home in cyberspace, which is

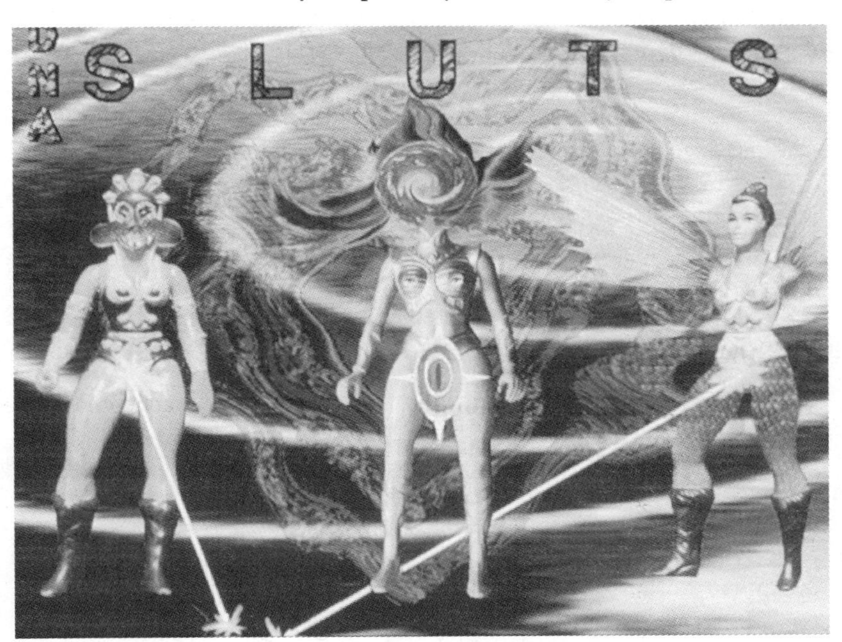

Figure 3: An example of the highly sexualized, aggressive cyber–femme fatale, as appropriated by *Wired* magazine. (The graphic accompanied Jane Szita's advertorial "Cyberdykes R Us," *Wired*, September 1995, 46.)

depicted as a pleasure dome for traditional heterosexual scenarios of (male) dominance and (female) submission.

Standard cyber–femme fatale discourse takes the form of a challenge or a seductive message laced with sexual innuendo and potential danger. For example, *Wired* describes an all female group of Quake computer game players that greets visitors to their Web site with the slogan, "Under every floral print dress lies a lady wearing black garters, carrying a big f*cking gun."[30] Similarly, one of Prodigy's advertisements for Internet service shows a dishevelled, long-haired beauty leaning over a car with a seductive, challenging expression. The caption reads, "Let's just say I don't hang out at the knitting forum."[31] The suggestion that the ad's subject, Loni, is available for online sexual encounters is clear. These aggressive warrior women of cyberspace are defended in *Wired* by the presentation of characters like St. Jude, portrayed as a postfeminist woman who dismisses feminist critiques of technology and argues instead that "tech will solve all our problems, personal and scientific. Girls need modems."[32] Such cyber–femme fatales are thus shown to reinforce *Wired* political discourse by echoing its libertarianism and faith in technological progress. These images are among the few found in *Wired* that seem to suggest women can be a part of technotopia. Judging by the images of women presented, however, they may only belong if they are willing to return to an idealized past of subordination (the cyber-Barbie), become sexy warriors for the new ideology (cyber–femme fatales), or (ultimately) strap on a VR device and leave their body on the other side of the keyboard.

Wired's exclusion and degradation of individuals and groups that do not fit the demographics of the digital generation is also evident with respect to racial minorities. Even fewer people of colour are depicted in *Wired* than women. This may be the result of the sexual interest brought by female images in the presumed heterosexual, masculine culture of *Wired* magazine. In any case, only one person of colour has appeared on the cover of *Wired*. On the December 1994 issue, the computer hacker John Lee is featured looking cocky, streetwise and in typical grunge style above a headline that reads, "Hacker Showdown: A member of a rival phreaker gang called John Lee a 'nigger' and in the hacker underground nothing was the same again."[33] His relevance to the digital generation is constrained by his status as an exception, within a racist western culture. He is not part of the techno-corporate sector; rather, he exists on the fringes, in the mysterious, shady world of the hacker, where mainstream culture would expect to find the stereotypic African-American gang member. His race, like Laurie Anderson and Sherry Turkle's gender, is both the reason for his inclusion in the magazine and the reason he is exceptional. Individuals who vary from the *Wired* demographic on more than one count are generally too different to be processed at all. Women of colour, for example, are extremely rare in the *Wired* world. On the few occasions when they are visible, as in the short 1998 article about Hong Kong born Aldeon Networks manager Selina Lo, their status in the high tech world is questioned and their credibility undermined. Lo's lack of "management experience" and "technical chops" are carefully pointed out, while her success is attributed to her "incredible knack" for product placement. She is also identified as an exception by the fact that she is not too scared to compete with other firms.[34]

The stereotypic constructions of the African-American male as gang member, rap musician or civil rights activist, are clearly the dominant constructions among the very few depictions of non-white men found in *Wired*. When articles do appear they marginalize and exceptionalize men of colour from the *Wired* world. Take for example the coverage of John Lee in "Gang War in Cyberspace," or "RapDotCom" that describes "how black people are giving the digital revolution the funk it so badly needs,"[35] or "Afroamerica Online" that reports on the "mission" of Malcolm CasSelle to "digitize, archive, and distribute Afrocentric culture in cyberspace."[36] A Lucent Technologies ad showing a business-suited Black man notes that "Everybody's getting hip to the power of K56 flex modem technology."[37] Because they represent the only images and text that reflects Black men to *Wired* readership, the effect is pronounced and decidedly racist. According to *Wired* magazine, digital culture, is largely, if not entirely, the purview of the white male. It seems minority figures are of interest to *Wired* only as tokens and stereotypes that reinforce assumptions of racial inequality.

Virtual Dreams of the Future: The Technological Elimination of Difference

The construction of an idealized digital Nirvana beyond the information revolution lies at the heart of digital discourse. Though *Wired*'s version is a more intense form than that found in the more traditional mass media, it is echoed in the images of mainstream computer corporation advertisements. Television commercials for high-technology services and products, for example, are often characterized by images of a future, utopian world of interactive communication across vast distances. Visions of *Wired*'s technotopia are lavishly exhibited in the regular four-page double-page spread at the beginning of each issue. These short sections, which lie between the first few pages of advertisements and the table of contents, provide a showcase of graphic design à la *Wired.* Whether it be a scene reminiscent of a high-tech CD-ROM computer game, a chemical world of constantly moving atoms, an elaborate pictorial flowchart, a series of digitally altered photographs or a montage of spliced-together images, this four-page spread gives the reader a glimpse into the hypermodern future envisioned by *Wired.* A nominal amount of text (a sentence or two) is superimposed on the visual presentation of these two-page spreads, thematically uniting the graphics. The text and image introduce the reader to the theme of one of the subsequent articles through a short direct quotation, frequently highlighting the most provocative elements of the text to follow. Though these spreads differ from issue to issue, they consistently combine provocative text, image and colour with a sense of high speed and movement. Text is laid out in unusual ways, forcing the reader to follow sentences horizontally, vertically or even in a circular direction. Though somewhat disorienting, the effect is dramatic and entices the reader to look for the article quoted in the way that an ad entices a consumer to buy a product. This elaborate technique, featuring minimal text and a glossy image, seeks to affect readers emotionally and transmits concepts visually rather than textually.[38]

These double-page spreads also have a more general function, beyond the promotion of a single article. The themes depicted introduce significant elements of *Wired* ideology. Readers are told, for example, that the digital generation has all the traits necessary to take power in the future, that the information age will decentralize authority, moving it away from the nation–state, or that future technologies will make society so efficient that people will live their entire lives in leisure. Such messages are dramatized by the complex, symbolically driven graphics that accompany them. The result is a powerful rendering of how the digital generation sees the future and what principles will guide its creation. Simply stated, the messages put forward the view that the digital generation will lead industrialized society into a new technotopic era in which online communities will replace the dying institutions of our present industrial culture.

But what sort of communities will they be? As we have seen, *Wired*'s technotopia often includes the reconstruction of traditional gender and racial stereotypes that would be better relegated to a distant past. However, *Wired* also frequently engages in utopian fantasies of a different, profoundly inhuman sort. Through the techniques of digital alteration and computer-generation, the images of people within *Wired* frequently take on a hyper-real appearance. Human images are often enhanced by an unfamiliar colour palate, fused with a computer-generated graphic or seen through, or surrounded by, computer screens. As one might expect there are also a number of photographs depicting human subjects wearing virtual reality headgear or other computerized attachments. Sometimes, the human form is fragmented and integrated within a larger computer system. The November 1995 cover, which depicts Nicholas Negroponte's face seen through a myriad of uniform pixels of light and colour, is an excellent example. Negroponte's distorted image is recognizable only as a part of a larger technological display. He is graphically represented as part of the technology that he discusses. Meanwhile, lighting, infused with neon or other equally intense colour, lends a futuristic look to many of the photographs. The resulting dehumanized effect reveals an important aspect of the *Wired* look: the human form becomes "wired" when mediated by high technology. It also indicates a significant aspect of *Wired*'s ideology, by suggesting that humanity will be increasingly linked with, if not subsumed by, computer technology and telecommunications in the future.

As we have seen, the creation of the *Wired* world as separate and distinct from the real world is reinforced by images of the natural world that are triumphantly synthetic. Though images of the earth are rare in this future-oriented *Wired* world, those that are presented are either idealized and digitally enhanced, almost to the point of nonrecognition, or are symbolically represented as an image of the globe hurtling through space. Frequently, the former depends on the use of digitally altered colour and texture to create hyper-real images that suggest rather than reproduce nature. These images, which may or may not be created from an actual photograph, are easily recognizable to computer enthusiasts as having been generated by computer-mediated publishing tools. In this way, computer technology is shown to create a "new and improved" synthetic version of the natural world. What is most alarming is that it is this synthetic version that is favoured most by *Wired* discourse. One of Packard Bell's computer ads expresses this preference by contrasting real-life experience in nature, complete with "malaria, venomous spiders and ankle-biting pygmies," with the wonders of video conferencing. It then asks its upper-middle-class white male readers, "Wouldn't you rather be at home?"[39] Apparently, high technology will make staying at home and experiencing the world as a simulation a welcome option for western men of the future. Meanwhile, images of the earth as a globe moving through space reinforce the male gaze in its most extreme and arrogant form: even the earth itself is not invulnerable to the knowledge and control of the technologically equipped "man." This objectification of the earth through visual representations, such as the globe, encourages the disassociation of humanity from the natural world, a philosophical precept that underlies much of digital ideology. Why trouble ourselves with the messy real world when technology can be used to discipline that world to meet our specifications?

Wired's coverage of the O. J. Simpson trial is a good illustration of how *Wired* ideology positions race and gender in relation to this future technotopia and provides an excellent example of how digital discourse uses high technology to eliminate the significance of difference. On the cover of its September 1995 issue is a digitally altered photograph of O. J. Simpson — a white O. J. Simpson. Next to it, and well before the innocent verdict was decided, is the headline "Innocent. Objectivity is Obsolete." The accompanying article contains another altered photograph, this time of Nicole Brown Simpson as a Black woman, being led by the wrist into a neon-lit night spot by a white O. J. On the facing page, the word "Guilty" is highlighted. The racially charged Simpson trial, however, is simply the hook into a *Wired* perspective on technologically induced social change. The real issue in a *Wired* world is not the race, gender or class of real people and how these aspects influence social interactions, but the ability of technology to manipulate such categories and effectively make them disappear. The digitally "enhanced" photographs of O. J. Simpson and Nicole Brown Simpson act as a metaphor for what technology can do for society — manipulate the categories to the point of meaninglessness and eliminate threatening social difference. This represents the ultimate *Wired* myth: the digital world of the future will overcome all the social conflicts of today by reasserting white male privilege in a disembodied virtual technotopia.

Another example of how technology can achieve such wonders is found in a May 1996 advertisement that features an African-American man in a contemplative pose behind a computer terminal. The ad's caption, which reads, "I marched on Washington and never left home," highlights the ability of Acer computer technology to assist the subject in his antiracist political work in a new digital world. Proclaimed to be "into progress. And empowerment," the ad's subject dons the clothes of the digital generation and uses the tools of white upper-middle-class men for his own work.[40] In this ad, the Other is melded into the self as all are immersed in the wonders of digital technology. Of course, in light of what we have seen of digital discourse and its repeated attempts to reinscribe unequal power relations of race and gender, it is unlikely that such an egalitarian future cyberworld is anything other than a myth. As Ziauddin Sardar perceptively notes, despite the rhetoric of prominent "netizens" about enhanced interracial interaction on the Internet, "the totalizing online character of cyberspace ensures that the marginalised stay marginalised."[41]

Much of "being digital" involves the relentless drive into enhanced virtual reality experience in cyberspace. It is this drive into a completely technologically mediated future that represents the (non)essential project of digital ideology and the all-encompassing dream of the digital generation. However, this drive is also underwritten by a palpable anxiety about embodied humanity and diverse identity. *Wired* ideology is incapable of equitably integrating collective difference into its construction of technotopia. Diverse identities, clearly recognizable to the xenophobic *Wired* homogeneous "community," are confined to exclusive constructions of the Other. Gender, race and class differences exist in *Wired* only in the presence of this terminably defined Other. In addition, as evidenced by the lack of images of women of colour, only one quality of difference can be recognized at a time. The mythology of cyberspace promises the destruction of such categories, the complete removal of different identities, such that individuals interact in an atmosphere devoid of race, class, gender and bodily existence. To enter virtual reality is to escape the identity politics of the real world, which to *Wired* is easily reducible to degrees of wireability. The end of identity politics in cyberspace is synonymous with the end of the Other as threatening difference. However, the exclusivity of digital discourse and the continued construction of technology as a male domain ensures that difference remains tightly controlled. In this virtual environment, stereotypic female characters can virtually exist in perfect security. Virtual women (or fembots),[42] do not, after all, threaten traditional gender roles or seek independence. Meanwhile, the disembodied hypermacho man is free to exercise his will in a digital free market that is directly and unequivocably under his control.

❖

Wired simultaneously emphasizes difference by presenting images and text that reinscribe hegemonic power relations and negates difference by excluding positive images of women and minorities and denying that digital culture is the creation of a particular dominant elite. In so doing, it presents a particular set of gender, race and class constructions that reflect an underlying ideology characterized by a strong belief in technological progress and the conservation of hegemonic power relations. Whether *Wired* is excluding, reconstructing or eliminating difference, women and minorities continue to be subordinated in the digital world it creates. Thus, although *Wired* comes wrapped in a dazzling, novel package, like much of the discourse of digital culture, it continues to sell a very old, all-too-familiar ideology: one that serves to perpetuate inequality.

Chapter Five

FILLING THE VOID: BUILDING THE HYPERMACHO MAN

> To escape his utter loneliness, his inability to relate meaningfully to nature or other cultures, even his own society, Western man seeks union with the only thing that he sees as redemptive: technology.
>
> — Ziauddin Sardar, "alt.civilizations.faq: Cyberspace as the Darker Side of the West"

> Shared knowledge connects the writer and the reader. It forms a bridge from the type on the page (or the screen) to the deeper meanings and nuances of words.
>
> — Constance Hale, *Wired Style: Principles of English Usage in the Digital Age*

THE INTENSITY WITH WHICH *Wired* discourse consistently excludes, degrades and attempts to destroy difference is equalled only by its almost obsessive reconstruction of white masculinity in a new, quintessentially hypermodern form. This new hypermodern machismo combines the mainstays of the emerging digital culture with very traditional constructions of masculine power, frontier mythology and technological transcendence. What emerges is an image of a hypermacho man who uses new forms of technology to reassert power. He dominates older forms of capital and the nation–state for the benefit of his new digital economic frontier, and reasserts traditional gender roles that protect masculine privilege. The *Wired* man is the creation of visual and linguistic devices combined with the long-standing symbolic associations that govern hegemonic western culture. The resulting discourse reaches out and tries to grab the reader personally, invoking images of hypermasculinity and supplementing these with very political, ideological content.

ATTRACTING THE RIGHT READERS

Although *Wired*'s design screams for attention, the high-tech intensity that characterizes the *Wired* look does not appeal to all potential readers. In fact, the style is uniquely designed to attract a particular brand of reader, namely, *Wired*'s self-proclaimed constituency and readership, the digital generation. It is this digital generation that is to find itself and a reality it understands graphically reflected in the magazine's design. As we have seen, other potential readers are often turned away at the first sight of *Wired*'s cover. As a result, the significance of the cover, as the primary identifier, cannot be underestimated. The basic elements of the cover do not stray from the conventions of commercial publications: a primary graphic features the character, author, personality or issue being highlighted in the magazine, and a series of headlines entice the reader below a consistent, easily-identifiable banner. What most dramatically distinguishes the cover of *Wired* on the newsstand is its colour intensity, the unusually provocative headlines and the creative, frequently computer-generated, cover image. Together, these elements create a strikingly unreal or hyper-real visual image that either attracts or repels potential readers.

In order to understand how this is done, it is useful to once again compare *Wired* with a more familiar and ubiquitous magazine genre, the women's fashion magazine. According to Ellen McCracken's useful study of this genre, the cover image of a model on a so-called women's magazine represents a "window to the future self," a symbol of what the reader can achieve by consuming the magazine's content.[1] The cover of *Wired* serves an analogous function as both the window to the individual reader's future and to a more generalized future world. The cover does more than simply catch the eye of the casual passerby. *Wired*'s cover graphic, which most often depicts a celebrity of the digital generation, challenges the (presumed) male reader to emulate the characteristics and achievements of the cover "model," who is almost always a white male. Just as the cover model of *Cosmopolitan* comes to signify the "Cosmo girl," and all the values endorsed by the magazine, so the figure on the cover of *Wired* represents elite members of the digital generation. And, like the model on the cover of a fashion magazine, the image on *Wired*'s cover plays on the vulnerabilities of its intended readers in order to draw them in. Female readers of fashion magazines find themselves drawn to the unrealistic, fantastic images of the current feminine ideal and their attendant promises of happiness and adoration; so the digital generation likewise sees on the cover of *Wired* a graphic representation of all that they (apparently) want to be. While the fashion magazine promises to replace anxiety and emptiness with the adulation that cosmetic beauty provides, *Wired* promises to replace a sense of lack of control or fear of emasculation with a reinvigorated form of masculine privilege in a digital world. How is this achieved?

Like the Cosmo girl, the *Wired* man is a figure who is mediated by the commercial products and themes that fill the magazine. Yet while female cover models wear the products that the fashion magazine advertises and project the flawless, current fashion image, *Wired*'s cover model is frequently depicted through a technological lens; he is shown superimposed on a surreal hypermodern background, or he is represented as a computer-enhanced image. He becomes a symbol of technology's wonders and a testament to the joys of a digital future. The *Wired* cover model does not need to wear any particular products, because he himself is infused with the most important product of all: technology. This association is enhanced by the fact that *Wired*'s cover models are usually leading technology gurus and futurists like Nicholas Negroponte or George Gilder, or computer game creators like John and Adrian Cormack and John Romero, or corporate leaders like Bell Atlantic's CEO Ray Smith, or Citicorp/Citibank's CEO Walter Wriston.[2] These figures are depicted as obvious masculine personas. However, unlike so-called cover girls, *Wired* cover icons look directly into the camera. They do not avert their eyes coyly or present a subordinate sex symbol image that invites the male gaze. Their stance is authoritative if not outwardly aggressive, and their facial expressions reveal self-confidence and power. Unlike the nameless Cosmo girl, the *Wired* man's name or professional affiliation is prominently displayed in a short headline that situates him as an important figure in digital culture. These headlines affirm the power and prestige of the featured figure, while connecting him to the *Wired* world. For example, Walter Wriston's image is accompanied by the headline, "He was the most powerful banker in the world. So why is he talking like a cyberpunk?"[3]

Rather than suggesting passivity and availability, the *Wired* man is definitely active, physically or intellectually flexing his muscles in a hypermodern *Wired* world. Despite the fact that his image is digitally altered or that he is flanked by an artificial background, he appears to remain in control of the image and the technology presented. He is not made attractive by "scientific advances in skin care" like the models in fashion magazines or patronized by hairspray's "bendy-holdy technology stuff" like models on TV ads. The viewer does not gaze at him but is confronted by him. As a result, the reader enters the *Wired* world as a challenge, not as a voyeur. Together, these elements reinforce a long-standing, exclusive cultural association that links technology with masculine power and privilege.

Of course, not all covers of *Wired* feature the elite members of the digital generation. The privilege has also been enjoyed by some more mainstream cultural celebrities. They too have been subject to the wonders of digital photographic manipulation and high-tech effects. For example, the race car driver Jacques Villeneuve becomes "wired" when he is represented in a highly stylized, technologically mediated image that links him to digital culture.[4] The fallen sports star O. J. Simpson is worthy of a *Wired* cover only when every aspect of his face (lips, nose, eyes) is technologically altered to portray him as a white man.[5] Even when *Wired* has no human subject on its cover at all, its technological preoccupation is clear as is its provocative, aggressive tone. Covers that feature cartoon images or bold graphics are less direct representations of the digital generation and more symbolic of the extent to which technologies, rather than people, are central to the *Wired* world. The accompanying headlines pitch stories in a dramatic way, frequently using war imagery or hyperbole to emphasize their significance. For example, one cover that generated controversy and resulted in a flurry of letters to the editor in the following issue features the Apple computer logo wrapped in barbed wire. Beneath it is a single word: "Pray."[6] The image emphasizes the technology-related content of the magazine and reflects the construction of corporate competition as an exciting, take-no-prisoners battle in which only the powerful will survive. Apple's shrinking market share, it would seem, makes it a candidate for divine intervention. The tag lines that accompany the Apple logo on this cover are typical *Wired* fare: "Starwave's Jocks Score," "The

Summer's Best Special Effects," "Telco Terrorism" and "Exclusive: Jacking into China." In these lines, technology combines with corporate power games and male sexual imagery to construct one of *Wired*'s dominant equations: technology = power = masculine privilege. Even when the face of the digital generation is absent, its interests and self-image are reflected on the cover of *Wired*.

THE HYPERMACHO MAN SPEAKS: THE SUBSTANCE BEHIND THE STYLE

Wired's hypermacho man is not just an image — once the image is created, he is given a particular language and political ideology. The language that is used works to exclude others from the discourse and helps to construct the ideological substance behind the hypermacho façade. Crucial to this ideology is the development of a very specific language that separates hypermacho discourse from the mainstream and codifies digital ideology.

One of the most remarkable things about this process is that *Wired* is candid about its aspirations for this novel digital discourse of the future. *Wired* discourse is not an accidental creation that has somehow taken on a life of its own. Rather, the editors of *Wired* have gone so far as to produce a handbook, *Wired Style*, that describes the "Principles of English Usage in the Digital Age." *Wired* sees itself as the creator of a new form of linguistic expression that breaks the rules of the past and sets new norms for the future. *Wired*'s hardcover style guide shamelessly implores writers of the digital age to "Be Elite," "Anticipate the Future," "Screw the Rules" and "Go Global." While the terms used in *Wired* may not be suitable in all contexts and for all audiences, *Wired* argues that "the underlying principles" of *Wired*'s trademark style still apply.[7] *Wired*'s editors go beyond simply offering definitions for the many new words and acronyms that have emerged in the growing computer culture; they are also involved in creating a new form of linguistic expression. This language plays a crucial role in how *Wired* creates the exclusive digital generation, the *Wired* world and, finally, the ideal hypermacho man. To crack the gender code of digital discourse, we must become familiar with this language.

The language used by *Wired* assumes a well-educated, well-informed audience with some degree of technical knowledge. As a result, articles make numerous unexplained references to literary, political, cultural and economic aspects of the past and present: they refer to everything from Lacanian linguistics to the popular Borg collective on Star Trek.[8] *Wired* articles also assume a basic knowledge of computer systems and a general understanding of the current debates and concerns surrounding digital technology. Far from having any pretense of inclusivity, *Wired* strives to be elite. The future is not for everyone; those who do not have the "cultural literacy" to understand the significance of such things as "the Clipper Chip" (a national data encryption standard), "Deep Blue" (IBM's massive chess-playing computer) or "Doom" (a so-called classic 3D-computer game), are of no concern to *Wired*. *Wired Style*'s editor, Constance Hale, describes this exclusivity as simply part of a larger cultural phenomenon that has seen the rise of niche markets and marketing. But *Wired* is clearly not performing the same task as a magazine for ski enthusiasts that describes the latest developments in ski equipment or recommends new resorts. *Wired* sees its role as far more expansive. While the skiers of the world have no illusions about changing the world, the digital generation sees itself as the "change leader" at the forefront of the digital revolution. Its exclusivity assumes an entirely different cast.

The members of this digital generation are not populist leaders who have risen from the grassroots. *Wired*, in the words of Hale, doesn't "dumb things down"; it believes in "going for the gonzo, the rough-edged, the over the top."[9] As a result, elevated vocabulary and complex sentence structures are used throughout. For example, simple words such as "use" or "improves" are replaced with more complex constructions such as "usage," "utilization" and "maximizes" or "optimizes." Grammatical structures often include a high ratio of syntactically complex sentences that express a series of logical relationships between pieces of information. This is particularly the case with regular columns like "Geek Page," which features such complex, technology-specific "explanations" as the following: "Both factorization and discrete logarithm problems forge strong cryptographic systems when they employ numbers that exceed 300 digits — or about 1,000 bits."[10] Computer and telecommunications advertisements in the magazine also assume knowledge about the components of computer equipment, as well as an understanding of acronyms. For example, readers are expected to understand what it means to have a computer with a 200-MHz Pentium processor, 32 MB RAM/3.2 Gigabyte HD and NEC MultiSpin 6 x 4 CD-ROM.[11] In fact, unexplained acronyms, which are pronounced as words (like the U.S. National Organization of Women, or NOW), or initialisms, which are simply clusters of letters (like the United Nations, or U.N.), are found throughout digital discourse. Of course, in *Wired*, most of these are specific to the high-tech sector and refer to all aspects of information technology, from the companies that dominate the field (IBM, MS, HP), to prevalent institutions (NSF, FCC), to more directly technological terms (ASCII, BASIC, CPU, DOS, LAN, HTML, PPP, SLIP).[12]

Further, all of *Wired*'s articles, regardless of their specific content, employ and develop the jargon of digital culture. Certainly, the inclusion of a regular feature devoted to the definition of new digital terms ("Jargon Watch") reveals the significance of such language to the editors of *Wired*. The terms that appear in "Jargon Watch" are more an indication of the *Wired* sense of humour, values and ideology than they are of words that are commonly used in the computer industry. For example, *Wired* defines "Cumdex" as the nickname for the pornographic technology showcase that featured items excluded from the Comdex trade show. An "MCI Project" is a low-budget "friends-and-family" sponsored endeavour.[13] "Survival Chic" is "a fashion trend centered around survival/emergency gear" and online gossip is dubbed "word of mouse."[14] Some, but not all, of the words introduced in "Jargon Watch" are commonly used in *Wired*. By littering these terms throughout the text, *Wired* includes elements that are familiar only to its regular readers and to members of the computer subcultures that now flourish on the Internet.

According to linguist Roger Fowler, the vocabulary of a language represents "a kind of lexical map of the preoccupations of a culture." Therefore, significant elements of that culture are heavily encoded in a language.[15] If this is the case, the digital discourse presented by *Wired* reveals that computer technology and telecommunications are the most significant elements of this emerging culture. The language used flags the text of *Wired* as separate and distinct and constructs the world of technology as a lawless, hostile frontier. Violent verbs such as "flame," "hit," "lurk" and "hack" describe online activities, while computer security systems are referred to as "firewalls," and the ultimate software application is a "killer app." Further, John A. Barry, author of *Technobabble*, notes that a high percentage of common computer terms were originally used to invoke male sexual imagery, whether or not they are still recognized as doing so.[16] Words used in all computer-oriented language such as "jack in" "joystick," "go down" and "head crash" are excellent examples. Such language functions to intimidate and exclude members of the larger society who are unfamiliar with these terms and who may find them violent or offensive. Indeed, this language acts as a gatekeeper to computer culture, subtly or not so subtly ensuring that digital technology remains the purview of the male sex. It also begins the work of constructing the digital hypermacho as the subject who is comfortable in the rough-and-ready competitive world of information technology.

Computer-related jargon is employed to describe much more than technological artifacts and processes, however. *Wired* makes extensive use of this jargon to describe "everything from human interaction to the state of the world."[17] Human relationships, indeed humans themselves, are described in terms of processes originally associated with computer systems. The result is a striking world with "scientists tweaking the boundaries of the unknown," everyone "interfacing" with everyone else and "hackers" "surfing the Net," while "P.O.N.A.'s, or People of No [e-mail] Account," get left hopelessly behind. Even the human mind is compared to a computer that "stores about a billion billion bits of information and runs at about ten million billion bits per second."[18] We will examine the philosophical and ideological implications of this language in more detail in the next chapter, but we should note here that the widespread use of such language creates a highly technologically determined, exclusive atmosphere. To enter the *Wired* world is to leave any non-technologically mediated, natural world behind.

Nevertheless, some elements of nature do find their way into the language of *Wired*. In fact, digital discourse frequently uses metaphors from the natural world to describe technology. Most often, this involves using animal names to describe technological tools in a kind of "wild kingdom of computers." For example, computers that aren't functioning properly are said to have "bugs" and "viruses." A well-known Internet protocol is called "gopher," an independently travelling destructive code is called a "worm" and a Web search device is a "spider." Digital discourse also makes reference to "hives," "the butterfly effect" and "cancel bunnies." These terms infuse technologies with lifelike qualities, thereby elevating their status from the mere collection of binary code or circuitry to living animals with a life of their own. As a result, those that work with computers are encouraged to think of various technologies as pets or independent living organisms, to be given at least the same consideration that would be afforded to the family cat. In this way, the language used by *Wired* simultaneously degrades what is it to be human, by metaphorically comparing human life with technology, and elevates technology, by bestowing it with independent, lifelike qualities.

This notion of technology as an active, even independent, agent is reinforced by the grammatical insights of Roger Fowler. Fowler suggests that the pattern of transitivity evident in a text indicates the dominant power relations the discourse attempts to reflect. That is, the way in which actions are organized in digital discourse offers clues as to the directions in which power flows in a *Wired* world. In the magazine, agency is predominantly given either to specific individuals, normally members of the privileged digital generation, or to technological devices that perform specific tasks. To Fowler, the dominance of such agent–action structures suggests a rational, instrumental world of "controlled activity."[19] However, human agency is sometimes obscured in order to subtly suggest technological agency and determinism. For example, *Wired* writer Greg Blonder describes programs that will help "the computer to anticipate changes in its environment," "sharpen its negotiation skills" and exponentially evolve beyond human capabilities. According to this article, "we will be driven to extinction by a smarter and more adaptable species — the computer." By arguing that "the computers of 2088 might not give us a second thought," Blonder further anthropomorphizes computers as thinking agents.[20] Meanwhile, technology is explicitly described by Jon Katz as an inexorable force that cannot be suppressed,[21] and Steven Levy refers to technological progress as the "inevitable trail leading to Digital Nirvana."[22] This message of technology-as-agent is reinforced by the common technical practice of using nouns as verbs in order to emphasize the power and excitement generated by technology. For example, computer enthusiasts are said to "network," while compatible systems "interface" and modems "handshake."

Amid this *Wired* world of active technology, people are depicted as fragmented, singular beings attempting to exert their influence in a hostile environment. Indeed, the "community" *Wired* constructs through its exclusive discourse and jargon does not form a collective, in the sense of a cohesive community. Rather, the community is a "wired" collection of atomistic individuals, each pursuing self-interested aims within the free market of cyberspace. While proponents of digital ideology, such as *Wired*'s executive editor Kevin Kelly, have also adopted the hive metaphor to suggest a future digital world in which a collective identity emerges through cyber-connection, individualism and highway metaphors still dominate *Wired* discourse.[23] This is evidenced not only by the highly individualized way in which *Wired* subjects are described, but also by the overt description of a very competitive atmosphere. For example, corporate chief executive officers are pitted against one another in "cable wars," and technologists strive to bring their product to market first.[24] The lack of communal atmosphere suggests an important site of tension in digital ideology. That is, while the digital generation pursues libertarian ethics that foreground personal freedom, privacy and lack of governmental regulation, they also purport to advocate notions of cyber-democracy and cyber-community. This apparent tension is resolved by the intervention of time. While the present process of technological change is constructed by *Wired* as a revolution, the futuristic digital world is portrayed as either an unexplored, wild and limitless frontier, or a transcendental realm of pleasure and salvation. Thus, although the present revolution shows little sign of democracy or community, technotopia will assure their ultimate realization. In the meantime, the digital generation, led by the hypermacho man, will guide society into that future.

THE TECHMAN COMETH: TECHNOLOGY + MASCULINITY = POWER?

Technology lies at the core of the idealized construction of the hypermacho man. By embracing technology, the hypermacho man gains access to the digital world and, like a comic book superhero, acquires his superhuman strength. In the words of one hypermacho figure, "I didn't do anything before I had a computer. Computers changed my life."[25] Technological fetishism — an obsession with technological artifacts — is a mainstay of *Wired*'s digital discourse. If readers are to emulate the digital ideal, they too must embrace the magic of information technology products. It is useful to conceptualize the pieces that perform this function in *Wired* as advertorials — a combination of an advertisement and an editorial. These advertorials can then be placed on a continuum. Advertisements that overtly sell a specific product are found on one end, and articles descriptively promoting a whole branch of science and technology on the other. The distinguishing feature of all these articles is not limited to their technological content, but also includes their continual obsession with the novel, their acute need for commodification and their overwhelming commercialism. Through these advertorials technology is raised to the level of the omnipotent. This is achieved through a number of common advertising techniques, including the rampant use of superlatives, wildly enthusiastic testimonials from users and graphic depictions of the product's awesome power. In addition, just as religious and war metaphors are used to elevate scientists and businessmen to the status of gods and conquerors in *Wired*, so are these metaphors used to sell the products of digital technology. Numerous articles focusing on specific technological developments are cast in the light of either of these dominant metaphoric forms. For example, a television camera becomes "the unblinking eye of God,"[26] and "a blue laser is the holy grail."[27] Computer hackers are engaged in "electronic gang war."[28] Many technologies are associated with the military–industrial complex and described in terms of warfare possibilities. For example, articles such as "Cyber-Deterrence" and "Warrior in the Age of Intelligent Machines"[29] introduce digital technology into warfare, highlighting its transformative and awesome capabilities.

These articles are reminiscent of mainstream newspaper stories that emphasize the potential of outer space as a battlefield through their descriptions of military initiatives such as the U.S. Star Wars project. In both types of articles, the location and technology of war shifts from conventional warfare on earth to technological warfare in outer space and, now, in cyberspace. The association of technology with military success and with spiritual salvation implicitly promise *Wired* readers an invigorated sense of power and strength. There is nothing, according to these advertorials, that technology cannot deliver.

In addition to these promises of glory, some of the articles and ads in *Wired* play directly on the vulnerabilites and insecurities of its readers and on their presumed desire to keep up with the digital generation and regain a sense of control in a period of rapid change. A Sony PC advertisement, for example, promises that by owning a Sony PC, "You will radiate confidence. Friends will ask if you got a new haircut. Associates will ask if you've lost a few pounds," and "relatives will ask if you're seeing someone new."[30] The accompanying graphic shows a business-suited white man walking on water, like some sort of technologically dependent Christ figure, with the help of his Sony PC. Lotus uses a more direct approach, telling readers to buy their software because "in this bloodthirsty business arena, it's be in control of your information flow or get crushed under it."[31] Meanwhile, articles like Tim Barkow's "Bottom Feeders" add credibility to such threats by describing how one small computer business was "eaten alive" in the "scrappy world of Silicon Valley."[32]

Another important way in which *Wired* elevates the status of the technology it sells and impresses upon readers its capacity to change their lives is through the creation of a digital history. For example, articles such as "Insanely Great: Ode to an Artifact" are a significant tool for constructing the technology of the recent past as an important element of social development. By arguing that the 1984 Macintosh computer "changed everything" and "began to weave itself into the fabric of everyday life," *Wired* exaggerates the all-encompassing significance of the device, well known to readers of the magazine. The accompanying graphic shows a Macintosh computer superimposed on an image of an electric explosion, producing a dramatic impact.[33] This reinforces the importance of contemporary technology and underlines the need for consumers, especially the "wired," to constantly upgrade and expand their technological resources. If readers are to gain some of the power and strength of the hypermacho man, they must use his history-changing technology.

When *Wired* discourse is not touting the magic of the technologies of the present or reminiscing about the successes of the past, it is selling the dreams of the future technotopia. In fact, the future provides a seemingly inexhaustible source of interest for the hypermacho man and his acute case of technolust. Future-oriented articles such as "Dream Ware," noting the extent to which prototypes define not only the future but also the present,[34] and advertisements such as Blue Sky Entertainment's CD-ROM promotion, welcoming consumers to the future of "interactive experience,[35] link dreams of a future technotopia to the consumption of technological products in the present. In fact, the future and the present are frequently conflated, intensifying the drive into increased digital commodification. The future is also objectified through the use of spatial metaphors which are sometimes combined with metaphors of discovery and adventure. Through the consumption and use of high-technology products, readers are promised a place in the future as they "explore cyberspace" or the "electronic frontier" in search of "digital Nirvana." This imagery not only injects excitement into the sedentary experience of clicking a computer mouse but also invokes common masculine frontier mythologies. The future, it would seem, is something "out there" that hypermacho explorers can find and conquer by using high-tech consumer products.

The construction of the hypermacho man is further enhanced with the use of digital generation "success" stories. These interviews or articles give individual accounts of inventors, scientists and businessmen who are on the cutting edge of technological development and marketing. The piece often begins with a highly personal description of the individual and develops into an increasingly technical and philosophical discussion of the related technology or research. His career history and an exhaustive list of his scientific degrees or leadership positions in successful corporate ventures are highlighted. In the dominant culture, these are the hallmarks of status and privilege. Combined with his high earning capacity, these attributes make the subject of the interview a "success story" of the digital age, someone who wields all the influence associated with wealth, power and wisdom. The conclusion muses about the future socioeconomic benefits that will be derived from further development of his work. All told, it would seem that *Wired*'s hypermacho idols gain their status through their association with science and technology rather than through their academic or business credentials alone.

These articles use hyperbolic, catchy headlines to create first impressions — "The King of Quant," "Revolutionary Evolutionist" or "The Father of the Web"[36] — and the photographs of the man himself are technologically enhanced to suggest power, authority and competence. The authority of hypermacho men is also created by using religious imagery. For example, Marshall McLuhan is called "Saint Marshall" and "The Patron Saint of *Wired*,"[37] while the early developers of the Internet are "The Great Creators."[38] Accompanying images and fonts invoke the celestial. And the interviewer/author rarely questions the technological expertise or ideological assumptions promoted by the subject, treating him as if he were a god and therefore infallible.

Despite their god-like status, these techno-heroes are portrayed as "just one of the guys." When dealing with elite members of the scientific and economic community, this is no small task. War imagery and metaphor, frequently used to emphasize masculine power and authority, are juxtaposed with the man's scientific or economic occupations to create humour and spectacle. For instance, *Wired* describes Tele-Communications Inc.'s CEO John Malone as the "Infobahn Warrior," Bell Atlantic's CEO Ray Smith as "The Cable Slayer" and Microsoft's mogul Bill Gates as the "Angriest Guy in All of Cyberland."[39] Below the headline is a digitally altered photograph of the subject himself, often costumed in the appropriate attire and situated in an appropriate place to suit the role. For example, the face of Bell Atlantic's CEO is superimposed on a photo of a figure that recalls Conan the Barbarian, complete with bulging biceps and sword. Similarly bizarre is the altered image of businessman John Malone, dressed in black leather and standing in a threatening pose on an empty highway, his faithful dog at his side, looking like a character from the 1982 movie *Blade Runner*. However, even when the depictions are humorous, a link is still made between the masculine and the digital, between power and technology. Humour simply makes the link less obviously offensive.

The articles also bring out the personal side of the man. They allude to a CEO's favourite actress, a *Wired* magician's quirky surroundings and pink desk, the whimsical T-shirt of a scientific genius.[40] In addition, the author or interviewer uses a style that renders the scientific and the financial entertaining. Technology and complex technical information is put in common digital culture metaphors, jargon and humorous narrative. In short, *Wired* attempts to make computer geeks and business-suited economic elites cool, exciting and identifiable to its predominantly male audience by emphasizing their significance to the digital revolution, their conformity to traditional masculine stereotypes and their status. This task is also achieved by the way in which the author or interviewer constructs his or her relationship to the subject. As a result, the subject becomes a celebrity of *Wired* culture, a hypermacho man to be idolized and emulated.

Another technique used to build the hypermacho identity is through the personal narrative. The author presents himself as a connoisseur of digital culture, the quintessentially wired man exploring the new digital frontier. Characterized by a first-person account of the author's experiences with various kinds of technologies, regular writers of *Wired* develop their personal authority and strengthen ties with the magazine's readership. These articles enhance the credibility of digital culture by providing first-hand testimonials to its many pleasures. It is within such personal narratives that issues such as cybersexuality and identity on the Net are addressed. For example, Dorion Sagan's article "Sex, Lies and Cyberspace" describes, in a conversational tone, his experiences flirting and gender-swapping on the Internet.[41] Other stories such as Joshua Quittner's "Automata Non Grata" and Robert Rossney's "Doom from Above" tell of adventures fraught with danger and excitement.[42] Success is ensured through technological knowledge, practice and courage. The fact that these exploits frequently involve little more than the manipulation of a mouse or joystick is mystified by descriptions that build on the cyberspace metaphor by suggesting the entry of the narrator into a completely different world.

These narratives are reinforced by ads that offer the reader the technological tools that will allow him to emulate the adventures described by *Wired*'s hypermacho writers. These ads, particularly those for CD-ROM computer games, entice the reader with highly visual, futuristic layouts and hypermacho constructions that promise violence, sexual excitement and heroism. For example, Paramount Interactive's Lunicus advertisement appeals to traditional myths of man-as-protector, forced to "waste" the enemy fast, lest they "kill [his] family,"[43] while Netrunner invites readers to "Jack into the Info War."[44] The 3D-computer character Duke Nukem is recommended to potential players as "Probably the only guy to oppose the death penalty on the grounds that it's more fun to do it yourself."[45] The complementary myth of man-as-sexual-predator is also used in a similar ad for Noctropolis, a game produced by Electronic Arts. Here, the hypermacho reader is invited to explore a futuristic world that unites sex and violence through "twisted desire" and "menace and allure" with "characters, devious and debauched." The "adventure in depravity" is graphically represented by a dark, shady world inhabited by stereotypic blond, highly sexualized women, clad in leather bustiers and chains.[46] Personal narratives describing hypermacho exploits and ads such as these identify the world of computer entertainment and the future playgrounds of cyberspace as male-dominated territories that cater to the whims of the hypermacho man. (See Figure 4.) The role of women in such a sphere is subordinate and often purely sexual.

Hypermacho Man Goes to War: Telling It Like It Is and Taking On All Enemies

Not surprisingly, in addition to being confident and aggressive, *Wired*'s hypermacho man is opinionated. In order to identify with the hypermacho man, the reader must also identify with those views that are closest to his heart. The two main issues that dominate this hyperheart? The protection and enhancement of so-called digital freedom and the future of the Internet. These issues are addressed with a strongly argumentative tone in articles designed to convince readers of their validity. They are characterized by common aggressive argument techniques. Like most aspects of *Wired*, these techniques are presented in an exaggerated form. The articles often have a caustic tone and make use of sarcasm and sweeping generalizations. Certainly, these articles are the most aggressive and decisive of the magazine. Raising awareness

Figure 4: The popular computer game character Duke Nukem: a quintessential hypermacho man.

about the protection of "freedom" on the Internet has emerged as one of *Wired*'s most important tasks. However, this issue is not presented as a complex debate involving diverse social groups. Rather, it is most frequently raised through the regular "Cyber Rights Now" feature, usually written by Brock N. Meeks, a strong supporter of the Electronic

Frontier Foundation. Dedicated to opposing "a few thousand voters who like running roughshod over the First Amendment,"[47] the series of short articles are introduced by such headlines as "Net Backlash = Fear of Freedom"[48] and are written with stinging sarcasm in order to win readers' solidarity and to discredit the opposition. Little in-depth information about the subject matter is provided. Instead, opponents (frequently congressmen) are reduced to caricatures and the argument about regulating Internet content is reduced to two options: either the Net is free or it is tyrannically controlled by a totalitarian regime intent on censoring public discourse. For example, Meeks describes the proposed *Communications Decency Act* as a bill that is likely to "cast a bone-deep chill across all forms of on-line communication, reducing them to nothing more thought-provoking than a Hallmark greeting card." He urges readers to "act now. And remember this: accept no compromises."[49] A tone of incredulity is often used when describing the opposition's position, which is presented as the simplistic rant of "barbarians," "technophobes" and "thought police." Hyperbole and humour are also used in order to make the point. For example, government regulation is purported to "water down on-line content until it becomes the intellectual equivalent of uncooked tofu"[50] and to be a part of a "wave of repression"[51] of the ilk found in George Orwell's *1984*. Readers are urged to take personal action in order to fight for so-called electronic freedom and make the Net safe for democracy. If they do, a completely deregulated telecommunications sector, like Bob Johnston's "Godzone" in New Zealand, will be the reward.[52] Such aggressive rhetoric and fear tactics indicate *Wired* has a strong ideological stake in this issue.

Wired's overtly ideological and philosophical articles also deal extensively with the future of the Net. These pieces are generally more subtle and the arguments more fully developed than in those found in "Cyber Rights Now." They are distinguished by an outward appearance of contemplation, a sense that the writer has not yet fully committed to the ideological position proposed. Articles such as "Is Government Obsolete?" suggest that the writer is grappling with the issue personally.[53] Yet because the issue is raised through a lens of ideological assumptions of positivistic, technological determinism and free-market economics, the conclusion is evident before the discussion has even begun. A similar dynamic occurs when issues anticipated to arise in the future are addressed. Here, too, issues are starkly presented as extremes, with one position clearly advocated. Concerns about the future of new information and communications technologies are commonly refuted by a final appeal to improved technology or by the argument that a greater degree of technological saturation is the key to ameliorating any future problems. By following *Wired*'s perspectives on these issues, the reader becomes progressively a part of the digital generation. He comes to identify with images of the hypermacho man and, in so doing, becomes familiar with *Wired* magazine's ideological positions. If *Wired*'s articles are sufficiently successful, he will soon find himself, in the words of Nicholas Negroponte, "being digital."

In view of the harsh, aggressive elements of the hypermacho construct and the digital frontier, it is not surprising that *Wired* shows little patience or kindness when addressing its critics. However, it does occasionally provide a forum in which to discuss detractors of the digital revolution and their dissenting opinions. The way in which *Wired* elevates its celebrities to the status of powerful victors and demigods is not repeated in the characterization of digital culture's critics. Two good examples are the articles "Digital Refusnik" and "Interview with the Luddite." The detractors are associated with Russian Communists and antitechnology rebels, so their dissenting voices are discredited even before any arguments are heard. The accompanying text and graphics suggest that the critics are psychologically unstable or simply ignorant. For example, Kirkpatrick Sale (a self-described neo-Luddite) is photographed smashing a computer, which suggests he is violent and unbalanced.[54] Sven Birkerts, the "digital refusnik," is shown in a photograph that blends his image with that of a manual typewriter.[55] Sale is aggressively attacked by the interviewer for spouting "nonsense" and being a hypercritical fanatic. Birkerts is treated more civilly but said to be "almost embarrassed by the anti-digital-firebrand role for which he is becoming known." In both cases, the dissenters are presented by *Wired*'s interviewers as being out of touch with digital culture and unaware of the full potential and inevitability of technological progress.

Moreover, dissenters' arguments are not presented by the dissenters themselves but through articles written as a response to their criticisms. The dissenter's view is paraphrased in a derogatory, simplistic way and discredited as being "one-sided" and "shallowly supported."[56] Such articles help to bring *Wired* readers together as they struggle against the ignorance of the nondigital. For example, the widespread concern that digital technology will result in increasingly dehumanized interpersonal relations is caricatured as "the chilling What-Kind-of-Country-Will-It-Be-When-Everybody-Is-Staring-into-a-Screen-All-Day-Instead-of-Touching-and-Meeting-One-Another fear."[57] Statistics are used to support the pro-technology views of *Wired* in one paragraph and maligned in the next when they appear to support antitechnology positions. Furthermore, all critiques of digital culture are labelled "Luddite" and dismissed as naive and futile due to their inability to accept the inevitability of technological "progress."

To strengthen their pro-technology stance, *Wired* also discredits the position of feminists and other social groups that have raised concerns about inequitable access to digital technology and pornographic and sexist material on the Internet. Feminist analysis is given short shrift in *Wired*— issues related to security and safety that are of particular concern to women, including the proliferation of online pornography, continually take a backseat to strident, anticensorship rhetoric. While the antifeminist rhetoric of academic Camille Paglia and the pro-technology cyberfeminism of St. Jude earn these women interviews in the magazine,[58] more critical theorists are vilified as "firebrand feminists" who fail to fully grasp the issues they address.[59] Opposition to a completely unregulated telecommunications industry is caricatured as a minority, extremist position. For example, Catharine MacKinnon is dubbed a "firebrand feminist" because she advocates regulation and so-called censorship, something that *Wired* suggests could severely disadvantage those MacKinnon should be supporting: "artists and lesbians."[60] The fact that the main target of such censorship is the lucrative (adult and child) pornography industry is thus obscured by digital discourse. Condemnations such as this demonstrate *Wired*'s hostility to feminist critique and to challenges that digital culture reinforces traditional views of women and power relations.[61] In addition, corporations themselves take issue with feminist concerns. Digital Pictures has answered criticisms about the sexist content of their games by running an ad superimposed on a bright two-page image of the American flag and suggesting that only conservative old men could oppose Night Trap.[62] Yet clearly a game that invites players to interact with scantily clad teenage girls who are victimized by vampires in full motion video would raise questions for many.

THE ULTIMATE TECHNOLOGICAL TRANSCENDENCE?

As we saw in Chapter 4, *Wired* magazine, like much digital discourse, frequently makes forays into technotopia, taking the reader beyond the digital revolution and into the cybernetic dream of a completely *Wired* existence. The virtual reality environments the magazine describes allows the melding of mind and machine and represent the ultimate escape fantasy. They also reflect a yearning for disembodiment. This flight from the body is evident in the digitally altered images of human physicality, the frequent images of people wearing VR devices and the discourse that invites readers to "experience the thrills of gearing up without ever leaving the comfort and safety of your home."[63] Within cyberspace, *Wired* readers are promised freedom as well as control of their environment and, most importantly, an escape from a world of constant change, uncertainty and potentially threatened hegemony. This construction of a quintessentially noncommittal masculinity was anticipated in Barbara Ehrenreich's *Hearts of Men*, almost fifteen years ago. In North American culture, according to Ehrenreich, the dominant culture's depictions of masculinity reflect the desire for individual freedom and a "flight from commitment." When combined with a cultural discourse of backlash and a climate of rapid technological change, is it any wonder that the hypermacho man has emerged as the ideal *Wired* subject? He offers the dreams of a future technotopia without demanding the sacrifice of the power and privilege western man has enjoyed in the past. In him, the digital generation finds the ultimate otherworldly expression of what David Noble argues western man has been looking for for hundreds of years: technological salvation.

Chapter Six

CODING DIGITAL IDEOLOGY

Rarely in the debate about technology does the voice of the citizen emerge. Specialized discourse, reduced to elemental soundbites, determines policy to an unprecedented degree. A technologically illiterate culture, inebriated with the spinoffs of high technology, hardly pauses to address the issue of the impact this environment will have.

— Timothy Druckrey, "Introduction," *Culture on the Brink: Ideologies of Technology*

Selling computers is about selling power.

— Karen Coyle, "How Hard Can It Be?," *wired_women*

We have to learn that diversity is not a recipe for conflict or chaos, but is our only chance for a more sustainable and just future — in social, political, economic and environmental terms. It is our only means of survival.

— Vandana Shiva, *Biopiracy: The Plunder of Nature and Knowledge*

WHEN I FIRST SAW *Wired* magazine, like many women, I found its design, imagery and rhetoric offensive. As I began to examine it more closely, however, I realized there is more to *Wired* than the reassertion of these sexist and racist images. In fact, there seems to be *much* more to digital discourse as a whole than the exclusion and elimination of difference and the reconstruction of very stereotypical, rigidly defined images of virtual virgins, cyber-whores and hypermacho men. I began to see a link between what *Wired* is doing and saying and the widespread social and political changes demanded by the emergence of the digital information age. As in previous periods of rapid technological change, an entire ideology is being constructed to support the digital revolution and to subvert critical debate. While a very important part of this ideology is the construction of masculinity, femininity and otherness, the rhetoric of digital discourse also includes a number of dominant myths that together form what is emerging as a new digital ideology. This hypermodern ideology stifles our participation and relentlessly implores us to "get wired." It is also intimately connected to the economic and political process that is rapidly changing our world, and takes a position on how economic, social and political institutions should function in the future. As a result, it, too, has very significant implications for women.

As part of a feminist politics of anticipation, we need to identify the myths that are circulated in digital ideology and that are manifesting themselves in the material realities of our lives. By doing so, we can identify the discursive strategies that digital discourse continues to use as it attempts to construct our future. We also need to understand how these ideological myths complement the sexist and racist images found in *Wired*. This chapter provides a brief overview of six of the key myths of digital ideology and highlights some of their potential dangers, especially as they relate to women. It also explores links between this emerging ideology and existing western feminist approaches to the technologies of the digital age.

1. WORSHIPPING THE MACHINE: TECHNOLOGY AS ENLIGHTENED DESPOTISM

"Behold, a dent in the universe."

These are the words used by the *Wired* writer Steven Levy to describe the impact of the development of the Macintosh computer in 1984.[1] Such hyperbole is not unusual in the emerging digital culture. Indeed, as discussed in Chapter 5, the theme of limitless faith in technological progress and the association of technology with religious transcendence is one of digital ideology's most important myths. Guided by an unfaltering optimism in the future and a strong belief in technological determinism, the myth of technological progress flows through the very ink with which *Wired* is printed. It is characterized by the almost completely a-contextual use of the word "technology." Technology is not embedded in complex power relations, but is a transcendent force to which all things must respond and adapt. It is a reified, even deified, phenomenon. It would seem that technology in western culture has become nothing short of a god. And the worship of technology, a spellbinding religion.

The hypermacho men featured in *Wired* are the devout followers and prophets of this male-oriented digital religion. In this context, one technology specialist is praised as a "technopagan" a "good Wirehead" and "in all ways *Wired*." Sometimes, these followers of technology are themselves compared to gods. For example, one computer game creator goes so far as to use biblical terms: "sometimes late at night, after I had done something really cool, I would look down on my creation, and I would say, it is good."[2] Nevertheless, even when the prophets of digital technology are granted such status, it is to technology that they owe their power and privilege. There is little doubt that technology makes the man; the man does not make the technology. In fact, to have *Wired* tell it, technology is an inevitable force, a gift of western science that propels humanity ever forward. Technology is treated as a force that causes things to happen and gives "mankind" an improved lifestyle, more choices and, above all, intellectual and creative freedom. An excellent example of this is provided by Steven Levy's discussion of the development of the Macintosh computer, said to have "provided us [humanity] with our first glimpse of where we fit into the future."[3] *We* don't decide what our future will be, technology does. Even prototypes for future products are described as "vision things," which define human perception of the present and the future.[4]

The human, sociological and economic contexts that surround technological developments are almost always ignored in digital ideology. While *Wired* does include discussion about corporations, individual scientists, designers and the state, the magazine's ideology does not allow these to sully the image of technology as the product of "objective" scientific discovery. Inventors in *Wired* invent inevitable technologies; the inventors are a means to giving technology life, not an interested party in its construction. Digital ideology does not recognize that individual interests are reflected in the technology that humans create. The technology that is invented is necessarily progressive — or is simply not placed within a context of social costs and benefits. While technologies clearly do affect how we perceive ourselves and our world, digital ideology removes human agency from discussion altogether as it relentlessly worships the machine. One of the results of this is a tendency to degrade humanity and to express anxiety about technology's superiority over humanity. The a-contextuality of technology has become so accute and the anthropomorphization of computers so complete that one *Wired* writer actually "reports on the latest battle to determine the most human computer, even as he worries that he may be the least human human."[5] Another contributor suggests that in order to prevent human extinction, brought on by the supremacy of the computer, "our only hope is to try and accelerate human evolution with the aid of genetic engineering."[6] This anxiety about the future of humanity runs throughout digital ideology. And it has consequences.

The lack of direct human agency removes any sense of human responsibility from the debate. If technology happens in a social vacuum, it is easy to unhinge technology's development from human values and concerns. Technology becomes inevitable. It is, ironically, a force of nature, an omnipotent god. And if technology is a god, then "man" cannot be responsible for its actions. An excellent example of this is provided by Kristin Spence's 1995 discussion of DNA testing in "The Biggest Little Lab in the World." Though Spence suggests that "it's impossible to avoid the moral issues surrounding this technology," such concerns are in fact easily dismissed because, as her final sentence indicates, "regardless, it seems we have picked up a book that will prove impossible to put down."[7] Oliver Morton makes a similar argument, pointing out in his 1998 article that while genetic engineering may be "unnatural" it is also "inevitable."[8] There is simply no stopping reified "progress." These views are echoed by liberal feminist approaches that suggest that since the cyberworld is a given we must all simply adapt to fit into its logic. Those who seek to place human priorities before technological imperatives — the neo-Luddites, for example — are bound to fail, because, according to *Wired*, "neither technology nor the essential human desire for change can be suppressed."[9] Simply stated, "human evolution is now inextricably bound up with technological evolution."[10]

This intense faith in technological determinism is increasingly evident throughout western culture. Cyberfeminist Sadie Plant reproduces technological determinism in her own work by suggesting that new digital technologies have gone beyond "man's" control and will themselves lead to a cybernetic, cyberfeminist future. So why is this a problem? While there is clearly nothing wrong with the development and use of technology for the benefit of humanity (there is no doubt that technology in many cases has improved human life), there *is* something wrong when societies are forced to adapt to technologies with which they have had no democratic involvement. In western democracies, changes in government policies and institutions are subject to (at least some) public debate. Yet the widespread restructuring of our economy through technological change is seen as an issue of economic efficiency and progress — not political democracy. This was certainly evident in Canada when the federal government's Information Highway Advisory Council proved to be dominated by business interests. The Council's final report in 1995 simply restated the priorities of the computer and telecommunications industries, spouting familiar buzzwords such as "innovation" and "competitiveness" and leaving little room for debate or public consultation.[11] Although western culture holds the values of democracy in high esteem, at least rhetorically, new technologies are designed, used and proliferated with little or no public discussion. Removing technological change from political debate complements the decline of the welfare state and the "cult of impotence." Once again our democratic participation in some of the most significant issues of our times is denied to us.

Digital ideology offers the promise of a new form of "cyber-democracy" — a kind of digital replacement for genuine democratic involvement in the decision-making process surrounding the production, use and proliferation of digital technologies. With the growth of online communities, it is assumed that cyber-democracy will develop as individuals and groups make rules for themselves in the unregulated, free market of the Internet. The Internet is touted as a decentralizing technology that will allow diversity to flourish. Yet as we have seen, the promises of cyber-democracy seem destined to serve the needs of the elite digital generation and the corporations which sponsor them. Their construction of the Internet's cyberworld perpetuates existing inequalities in order to preserve their privileges. In a virtual community populated by those who log-on, what will democratic freedom mean? Digital ideology defines freedom as freedom from restraint, freedom to be anonymous in a virtual world of the future, and the freedom to avoid the claims of others. But what constitutes a real community and how will these virtual communities deal with issues of responsibility and accountability? In a world of anonymity where real identities are multiple and obscured by online pseudonyms can there be accountability? Can there be democracy without accountability? Is it possible for democracy to exist in the context of a virtual world that does not demand its citizens to commit to the welfare of others? How decentralized can the Internet really be when it is increasingly owned and operated by a small group of powerful multinational corporations? Without the participation of diverse people (many of whom are excluded from digital culture), cyber-democracy will pose little threat to the elite interests of the digital generation. But it will undoubtedly present challenges for the rest of us.

2. Cyberspace Won't Hurt the Real World

> No ambition, however extravagant. No fantasy, however outlandish, can any longer be dismissed as crazy or impossible. This is the age when you can finally do it all.
>
> — Ed Regis, "Meet the Extropians," *Wired*, October 1994

> All life, at its core, is a process of digital information transfer.
>
> — Michael Schrage, "Revolutionary Evolutionist," *Wired*, July 1995

A second significant feature of digital ideology is the creation of a mythical, almost divine, crusade into cyberspace. It is there, in what William Gibson has called "a consensual hallucination," that mind and body will allegedly fuse, that reality and virtuality will become indistinguishable and that space and time will yield to human desire.[12] Through the cyberspace metaphor technology becomes a means to attain a masculine, "otherworldly" transcendence. In cyberspace, a technologically determined heaven or digital Nirvana, inhabited by disembodied male computer wizards, becomes a very real possibility. In *Wired*'s words, the completely artificial, technological environment of cyberspace "may soon appear as strangely sentient as the forests in which the first magicians glimpsed the gods."[13] The myth of cyberspace provides a possible exodus for the privileged from a messy world of conflict. It is nothing short of regaining paradise lost, because in cyberspace, "you can create your own simulated universe if you want to."[14] The way in which cyberspace is constructed in geographical terms as a separate world — a virtual Garden of Eden, constrained only by the present limits of technology — colours all of digital discourse and ideology.[15] The fact that cyberspace actually depends on a very physical cable system that links physical spaces across time (albeit with remarkable immediacy) is mystified by the construction of a metaphysical Garden of Eden to which the computer screen acts only as a window. This construction glamourizes computer networking and perpetuates a myth based on the devaluation of the physical world in favour of a world completely mediated by technology. The degree to which this technology, by its very design, limits the scope of that possible "world" is denied.

If cyberspace is *Wired*'s new Garden of Eden, the natural world cannot help but pale in comparison. The alienating manner in which the natural world is portrayed in *Wired* suggests that digital culture is much less impressed with the earth's natural environment than it is with its own simulated one. Nature is fragmented, objectified and commodified through *Wired*'s editorials and ads. The globe itself is objectified, reduced to a toy, a plaything. Advertisements such as IBM's OS/2 Warp promotion uses the image of the globe as a backdrop to display its technological successes and the extent of its product's worldwide penetration.[16] TDK reduces the earth to digital bits of information by showing an image of the globe disappearing into a compact disc.[17] Meanwhile, to hooked.inc. the world is little more than a plaything, a globe/basketball held in the outstretched hand of the Internet user.[18] Worldwide Internet Publishing depicts the earth as an olive, pierced by a toothpick at the bottom of a martini glass.[19] This degradation of the natural environment through the manipulation of globe imagery is complemented by what John A. Barry describes as "the aggrandizement of the commonplace" through the overuse and misuse of the term "environment" in computer jargon.[20] Computer environments, such as the Internet, come to signify the entire world, as though nothing is lost or altered by digital "translation." Certainly, the significance of the term "environment" is lost when computer systems, interfaces, programs and desktops are all said to have their own environment.

The next step in the objectification of the organic is the exploitation and degradation of the environment that sustains it. According to the ideology of digital culture, nature simply provides resources and raw material for technological creation and manipulation. The logic of what *Wired* calls "deep technology," a twist on "deep ecology," suggests that "the more technology we have, the more we depend on nature, the more resources we need, and the more energy we must find to keep our intensive lifestyle afloat."[21] In *Wired*, the term environmentalism is reduced to the need to maintain the environment as an ever-ready supplier of silicon. In a February 1997 interview with business administration professor Julian Simon (also known to *Wired* as "The Doomslayer"), the fact that the natural environment is suffering from human exploitation is dismissed as being absolutely wrong. Despite abundant evidence to the contrary, the article insists that "the environment is increasingly healthy, with every prospect that this trend will continue."[22]

Environmental issues are absent from digital ideology. It is assumed that new technologies are benign because they do not evoke images of smokestacks emitting filth. This is not the case. One has only to consider the millions of dollars that have been spent on cleaning up contaminated groundwater in Silicon Valley, home of the computer industry in the United States, to see the ugly side of this rising industry.[23] The short life cycles and designed obsolescence of many digital technology products suggest that the industry sees itself as exempt from environmental action.[24] Yet the high use of plastics and toxic chemicals in the production of these technologies should raise serious concerns about the long-term sustainability of the digital information age. Everywhere microelectronics are manufactured researchers have found environmental problems, some of which have led to widespread health problems including high rates of birth defects and miscarriages. And what about the much vaunted paperless dividend of computer use? Far from creating a paperless office, more and more of the world's trees are needed to feed the computer printers of the new digital offices in pursuit of hard copy perfection. In addition, the proliferation of computers has contributed to an increase in energy demands in North America.[25] This is to say nothing of the radiation emitted by video display terminals. The fact is, far from the clean industry it is purported to be, the computer industry is responsible for multiple forms of pollution and for resource depletion worldwide. Unfortunately, liberal feminists and cyberfeminists also fail to acknowledge these significant environmental concerns. Perhaps, as the rhetoric of the digital ideology indicates, there is simply not time in the great haste to meet the needs of global competition to discuss long-term environmental effects. But there should be.

Despite *Wired*'s claims to the contrary, western culture has been repeatedly informed of the consequences of maintaining such an instrumental, reductionistic conception of the natural world.[26] Ordinary citizens constantly express concern about the state of the environment and show a willingness to help protect resources for the future. However, corporate interests remain focused on an even greater degree of commodification and exploitation. We have been responsible for the extinction of many species, for such phenomena as global warming, and for the destruction and erosion of life-giving forests and soils. Is it really possible to still dispute the reality that the earth cannot continue to sustain a population so bent on its own destruction?

3. No Time Like the Future: Novelty in a Hurry-Up-and-Wait World

So what is the new economy? ... A world so different its emergence can only be described as a revolution.

— John Browning and Spencer Reiss,
"Encyclopedia of the New Economy,"
Wired, March 1998

Have you ever installed a phone on your wrist? You will.

— AT & T advertisement,
Wired, March 1995

Change is good ... Time is the only truly scarce commodity.

— The editors and designers of *Wired*,
"Change is Good," *Wired*, January 1998

Digital ideology describes developments in new information and communications technologies in terms of another important myth: the myth that the current process of technological change is completely unprecedented, necessarily rapid and the way — the only way — of the future. As we have seen, there is much evidence to suggest that technological restructuring will continue to affect the lives of millions of workers and facilitate the flow of international capital with ever-increasing speed and ease. These effects also include the restructuring of corporate and social organizations that would ideally, according to *Wired*, result in "the elegant, streamlined, quick-as-a-keystroke style of work that computers have made possible."[27] It is important to recognize, however, that the language of change used by magazines like *Wired* obscures the fact that many of the existing dynamics of modern life will remain very similar in the bright "new" digital age of the present/future. Many of the changes we face are less revolutionary than they are an intensification of existing processes.

Digital ideology's concept of rapid, widespread change — likened to such momentous human events as the discovery of fire — does not include an end to social and economic injustices, inequality, the exploitation of labour, environmental degradation, unquestioned technological progress or faith in an "objective" science. In fact, western science and technology still hold a place of unquestioned authority in digital ideology, and American corporate and cultural hegemony continues to be a force to be reckoned with, despite claims of a growing multicultural online environment. And whether or not we are wired to the Web for hours each day, we will still have to eat. And sleep. And do a variety of things that involve the exchange of goods and services within the context of global capitalism. While digital ideology promotes novelty and social revolution, such claims are, like personal computer ads that purport to "change everything," highly exaggerated. Rather, a discourse of change is used to claim that the new economy is so unique and unprecedented that it is beyond our comprehension and beyond the control of existing government structures and methods of regulation. At the same time, digital discourse vigorously attempts to influence the future by reasserting very traditional sexist, racist and classist views. Like the Industrial Revolution before it, the information revolution seems to be reinstitutionalizing women's roles in regressive forms and redisciplining categories of social difference to serve the needs of the privileged. Once again, women are being moved out of the labour force and used as home data-entry personnel or as the disposable factory workers of the computer industry.

Meanwhile, digital ideology continually obscures the connections between current processes of change and past ones by presenting an ahistorical view of technology and associating the digital age with a conflated present/future.[28] As a result, digital ideology frequently claims that the "future is here." The fact that the future, by definition, will never actually arrive is crucial. By intimately connecting much of its discourse to an always unattainable future, digital ideology finds another way to subvert criticism of its numerous problems. Concern about widespread access to digital technology, for example, is described as a matter that will be inevitably solved through the future proliferation of communications technologies. As in the case of trickle-down economics, there is an assumption that with time, the wealth of the information age will filter down to all levels of society. In the meantime, we must all hurry to get online. There is no time for questions or open discussion.

The fact that there is no program in place to ensure equitable access to high technologies is ignored. Exactly when the benefits of new technologies will be concretely felt throughout society also remains unclear. And what exactly will these benefits be — beyond such questionable conveniences as online shopping and banking? As corporate dominance of the Internet increases, how valuable will access to online resources really be? If these services do become widely available, will the Internet's much-celebrated decentralization remain intact? In view of the increasing corporatization of online environments and the high-stakes battle between telephone and cable companies for control of the required infrastructure, it is likely that new Internet services will be useful primarily only to enhance our role as consumers. This is particularly the case for women, who remain underrepresented in positions of power in the world of new information technologies. The development of so-called push technology, as a means of sending customers only those services that marketing surveys indicate they want, is clearly a move in this consumer-oriented direction. Meanwhile, digital ideology uses uncertainty about the future as a way to deflect any criticism about these developments. By constructing the digital age as always just around the corner, digital ideology can forever defer responsibility for the future world it is creating.

If faster really is better in the digital age, we may well ask who benefits from such speed. If the hypermodern world of the future is constructed as one that will never actually arrive and yet one that demands that we all jump onboard as quickly as possible, there is little possibility of social criticism developing. Critiques of digital ideology will always be wrong, because, according to the digital generation, we are living in a world that is simply too fast and too future-oriented to understand in the present.

4. MONEY AND ME: DIGITAL REDUCTIONISM AND LIBERTARIANISM

> Money is just a type of information, a pattern that, once digitized, becomes subject to persistent programmatic hacking by the mathematically skilled. As the information of money swishes around the planet, it leaves in its wake a history of its flow, and if any of that complex flow can be anticipated, then the hacker who cracks the pattern will become a rich hacker.
>
> — Kevin Kelly, "Cracking Wall Street," *Wired*, July 1994

> Information plus technology equals power.
>
> — Hugh Gallagher, "The Accidental Zillionaire," *Wired*, August 1994

According to digital ideology, information is the lifeblood of the digital age. In *Wired*, the term "information" is grossly overused and signifies many things: data, bytes, graphics, text, even knowledge. The entire world, according to *Wired*, can be easily reduced to bits of information. It is important to recognize, however, that digital ideology operates with a particular concept of what information is and what purpose it has in the larger society. Digital ideology is largely incapable of effectively distinguishing between information and knowledge. While it would seem legitimate to advocate the widespread sharing of knowledge, digital ideology frequently reduces knowledge to information through a process of objectification and commodification. It is then possible for digital discourse to conflate not only knowledge (or meaning and experience) with information (or raw data), but to further reduce information to exchangeable wealth. For this reason it becomes not only possible, but inevitable, that living organisms become things that can be patented, commodified and sold. Life forms are treated like machines instead of self-organizing and self-reproducing organisms with intrinsic worth.[29] Digital ideology places all information — whether derived from artifice or nature — squarely within the economics of free-market capitalism. Why should DNA be any different? The natural and the "manmade" are basically the same according to the reductionistic views of capitalist entrepreneurs, as described in the article "Hacking the Mother Code." According to them, "the gene is by far the most sophisticated program around."[30] Complex natural processes are described as "codes" or "programs" that are "full of bugs" and are in need of "upgraded versions." And the mysteries of life? According to *Wired*, "life is no mystery — it's digital."[31]

Within this context, *Wired*'s proclamation that "information wants to be free" becomes somewhat ludicrous. In a future economic order in which information is a means of achieving wealth and power, "free" is the one thing that information can never be. If free information were actually the goal, securing intellectual property rights for corporations — what Vandana Shiva calls "enclosing the intellectual commons"— would not be a desirable goal. If securing profit were not a motive of the information economy, it would not be necessary to secure legal protection for digital investment flows or to secure the right to "own" patents to seeds and organisms.[32] Yet for *Wired*, freedom seems to be less about equity and reasonable cost than about removing all external restrictions. Thus the digital generation is opposed to any regulation of electronic information, whether it be for the purpose of privacy and security, for protection from hate literature and child pornography or for the regulation of capital mobility.

Wired's digital libertarianism is ironic. If there is one structure that is a testimony to the importance of government involvement, it is the digital information system, which owes its existence to a U.S. government military project. The significance of government involvement does not end there, however. Indeed, without government subsidization it is questionable whether the Internet would have survived at all, let alone grown.[33] It is also difficult to imagine how profits could be made in an information economy (or any economy) without the necessary government legislation to protect and facilitate capital accumulation. And who would design the hardware and software necessary for the telecommunications industry without government-subsidized educational institutions? While information technologies certainly facilitate such economic restructuring and speedy capital mobility, they do so within a legal framework that exists either inside national borders or that is negotiated within international organizations by national governments. Indeed, the fact that such legal agreements are still necessary for hypermodern capitalism to function is indicated by the members of the Organization for Economic Cooperation and Development rushing to secure a Multilateral Agreement on Investment. Similarly, the securing of Trade Related Intellectual Property Rights in the General Agreement on Tariffs and Trade was crucial to facilitating corporate biopiracy in the South.[34] Digital revolution or no, national governments and international agreements still have a role to play in this new economy. Yet despite the enormous contributions made by the public sector — by all of us — to the information technology industry, digital libertarians now demand that the state keep its "hands off." Is it a coincidence that digital libertarianism has arrived on the political spectrum just as the digital information industry has become so hugely profitable?[35]

As the information economy continues to develop and define the way commerce and business service will be conducted in the future, monopoly infrastructure companies are gaining more and more power to control the content and use of the Internet. In fact, digital discourse's promises of unlimited access to quality information appears very shaky in the context of corporate capital. While only a few years ago the Internet remained relatively free of advertisements, the online presence of corporations is now ubiquitous. It is becoming more difficult to find corporate-free zones and to wade through all the commercial material to find specific information. Without some sort of third-party regulator (a job usually ascribed to government), will quality alternative information services and resources become the online equivalent of the struggling publicly funded TVOntario in a television world dominated by corporate giants like TimeWarner? Multinational corporations clearly have high stakes in creating and controlling digital technology and its infrastructure. Whether or not Microsoft is successful in fending off antitrust lawsuits brought against it by the U.S. Justice Department, the company still holds the license of the computer industry's standard Windows operating system. Monopoly, it would seem, is already in place with regard to some aspects of digital technology and is likely to develop in others. Viable suggestions for protecting citizens' rights to use digital technology for nonexploitative purposes and to have genuine choices and access are absent from digital discourse. Thus far, digital discourse has failed to address issues of freedom beyond its opposition to government intervention. Rather, it has functioned to sell the public a vision of widespread technological change as inherently liberating and progressive.

Like any other commodity of capitalist accumulation, digital ideology markets information within a paradigm of inexhaustible need. It is assumed that the more information, the better; why live in a twenty channel universe, when two thousand are possible? Information is not differentiated in terms of quality but is viewed instead as quantitatively significant as a means to accumulate power and wealth.[36] Unfortunately, as we continue to be bombarded by what has been called a "data storm," it is more and more difficult to achieve shared understandings in the face of difference. We become overwhelmed by rapid-fire, dazzling information flows that leave little time for reflection and treat culture as just another commodity. Television failed to provide the widespread dissemination of quality information and diverse views it initially promised; and it is unlikely that the interactive "computainment" of the future will be able to deliver this type of information either. Online interactivity has not led to the majority of users learning about different cultures and views. Rather, social divisions, including racism, sexism and classism, are simply reinscribed in separate online environments that cater to the needs of different groups.[37] In addition, many Web surfers have found that online content does not live up to its reputation: "that half of cyberspace which is not commercial is largely 'toilet wall,'" a climate in which many are simply "projecting their darker side on the hypertext world of cyberspace."[38] While excellent resources are available online, these do not account for the majority of material that can be found on the Net.

Yet conventional wisdom tells us that in order to compete in the global marketplace of the future, computer literacy and access to as much information as possible are essential. As a result, departments of education around the world have invested heartily in outfitting classrooms with the latest computer technologies. An article in a recent issue of *Owl Magazine* tells children that "surfing the Net is where it's really at." It is accompanied by an image of a young white boy using a computer keyboard to surf a wave.[39] Once again, technology is visually depicted as a masculine enterprise linked to other macho activities like surfing. Many children's computer games also reinforce sexist stereotypes.

Girls continue to be socialized in a way that discourages them from engaging in computer-related activities and teachers report that even very young girls exhibit an aversion to computers that suggests they are well aware of cultural messages of technological exclusion.[40] But children are learning more than this from computers. Far from the pen-and-paper it is often compared to, the computer is not a neutral educational tool. Rather, computer technology amplifies the significance of information that can be reduced to bits and manipulated in a digital environment. Much of the context is missing. As a result, there is a greater tendency to obscure the extent to which knowledge is produced collectively and subjectively. So, are classrooms that are online better than those that aren't? Not necessarily. They may have access to a greater amount of information, but that does not mean that their students are learning more. As Theodore Roszak notes in his influential *The Cult of Information*, "the mind thinks with ideas, not with information."[41] If teaching children how to think (not what to think) is the goal, then bombarding them with information and showing them how to use software to manipulate the data is not the way to achieve it.

Furthermore, studies indicate that much educational software encourages both a high degree of individualism and an exaggerated sense of personal power, similar to that found in digital ideology.[42] This ideological content is no substitute for learning the skills required for genuine democratic involvement in society. It is debatable, then, whether software that encourages these characteristics will contribute to the growth of responsible citizens. There can be little doubt, however, that individualistic, autonomous students will make for good consumers, hungry for technology's latest "powerful" commodity. What does this suggest about the effects of software designed by the digital elite on the education system? Does this help to explain why software companies are so interested in providing computer services to children? And why governments, even in times of fiscal restraint, are so eager to deliver them? In a world full of rapidly changing social and political problems — one that demands greater communication and critical skills than ever before — we may not be doing our children any favours by putting them in front of a computer screen all day.

5. Imperialism Continued: This Time It's Digital

> What is at risk in the arguments over culture is France's — and Europe's — stake in the future. Any culture or nation that does not come to grips with technology is living in the past.
>
> — John Andrew, "Culture Wars," *Wired*, May 1995

> An essential component of the emerging global culture is the ability and freedom to connect — to anyone, anytime, anywhere, for anything.
>
> — Leila Conners, "Freedom to Connect," *Wired*, August 1997

Despite the myth that there is a diverse "global marketplace" that is accessible via the Internet, digital ideology retains significant elements of western imperialism. In fact, it seems that the globalization advocated by digital ideology involves less an exchange of cultural knowledge (or information) than a self-interested extraction and commodification of subordinate cultures. Western digital technology has no compunction about imposing its ideology in foreign contexts in an imperialistic manner. This extraction usually takes the form of cultural reduction and commodification based on racist western stereotypes. For example, in *Wired* discourse Africa is associated with music, the body and so-called fuzzy logic, and is seen as a source of cultural information to be utilized and refined by western artists and designers.[43] Indeed, like virtually everything else, art and culture are commodified by digital discourse as information bits to be traded across vast networks of fibre-optic cable. A lack of respect is shown for the integrity of the world's diverse cultures in digital ideology. The realities of the globalized computer industry are more realistically portrayed by the conditions found in factories that manufacture semiconductor chips in Malaysia. There, a factory manager for Motorola acknowledged the need to "change the culture" of Malaysian society in order to facilitate the exploitation of non-western women's labour.[44] The effects of foreign businesses and technologies on indigenous cultures are not considered relevant to the larger goal of capital accumulation. As has become clear from the modus operandi of institutions like the World Bank that operate in the interest of western industrial nations, a price must be paid for inclusion in the global marketplace of the future. Even if your place in that global marketplace is one of exploitation.

Meanwhile, the glaring whiteness of digital culture is reinvented as the white box of the Macintosh or PC computer and deployed as a civilizing force of the information age. TV advertisements such as IBM's "Solutions for a Small Planet" and Microsoft's "Where do you want to go today?" emphasize information technology's ability to reach the far corners of the earth, liberating and empowering all who receive it. Even one of *Wired*'s subscription advertisements, which superimposes the slogan "Get *Wired*" and its e-mail address over a black-and-white photo of a crowded street in a developing country, clearly advocates digitization as development.[45] The imperialistic aims of info-capital are masked in the warm glow of cyber-connection; the dreams of globalization are revisited in the form of ones and zeroes. *Wired*'s editorials and feature articles provide many testimonials to the fact that technology has extended its reach beyond the digital generation. There are stories of Tibetan Buddhists consigning ancient holy texts to cyberspace,[46] images of Powerbooks deep in the rainforest, and articles describing how "the resistance network ... gives hope to the oppressed around the world."[47] The profound impact of technological imperialism on other civilizations and, indeed, on every realm into which it enters is not addressed. Like the civilizing mission of the colonial period, new digital technologies are sold to the world as a new liberating "truth" in a crusade embarked upon in the name of development.

Digital ideology views opposition to technology's globalizing trends as regressive obstacles to be overcome in the pursuit of so-called progress. Not surprisingly, much of this opposition is attributed to "foreign" national governments that are ignorant of the exigencies of the digital age. According to *Wired*, the "Chinese were born to hack," but suffer from a backward communist past;[48] the French are fighting a losing cultural war against the dominance of the English world; cultural protectionism has "reduced Japan to a third-rate power in networking";[49] and "any culture or nation that does not come to grips with technology is living in the past."[50] Meanwhile, it is difficult to dispute the fact that American corporations dominate the new information and communications sector and that American culture exerts an enormous influence over global practices and beliefs. The dominance of Microsoft alone cannot be ignored: this one American company holds the license to the Windows operating system that is used by 90 percent of the world's computer systems.[51] A new digital world, founded on a global information system run by American-based computer corporations, would seem to be the unstated goal of digital ideology.[52]

The dangers of such digital imperialism for the future are not difficult to imagine. The most significant of these is the exploitation and disenfranchisement of a great number of the world's people. If reinforced by investment agreements like the proposed Multilateral Agreement on Investment, leading corporations of the new digital economy may assume they have a legally protected monopoly on the decision-making process that will affect the lives of millions of workers. Far from inviting diversity, it is most probable that digital imperialism will anglicize (or Americanize) global communications as a few transnational corporations divide up the spoils of the global economy. Caught in one of the most vulnerable positions in the digital economy are the many young women who endure the exploitative working conditions of factories that assemble computer equipment or enter data into computer networks. For them, digital technology does not offer a utopian vision of transcendence in cyberspace, but rather a continued struggle for immanence.

6. The Final Appeal: The Ironies of Production and Consumption

Because it's a FIRST. Because YOU want to be first. Because you want it NOW. Because now YOU CAN.

— Joint advertisement for Netscape and Visa, *Wired*, March 1995

Some dream of a perfect game. The rest of us just buy it.

— Hardball 4 computer game advertisement, *Wired*, January 1995

Patterns of western consumption play a significant role in the creation of the new digital age. Consumerism provides a way to spread digital technology throughout society, thereby ensuring its future dominance and creating dependency. *Wired* relishes its consumer impulses and flaunts its techno-lust through regular features such as "Fetish." This near-obsession with high-tech products, described by Marike Finlay as "gadget-philia," reveals more than an innocent fascination. The result is much more intense: When a product fascinates us *"it holds a certain compulsive rather than merely repressive power over us.*"[53] Such rampant high-tech, high-end consumerism is the domain of a privileged virtual class, able to keep up with the short life cycles of new information and communications technology products. It also reflects the West's industrial arrogance and the computer industry's willingness to turn a blind eye to the environmental effects of disposable technologies. Just as the fashion industry continually recreates what's in style, the computer industry continually reinvents what's "wired" and what's "tired" in the world of high technology. Upgrades are constant and delivered at a frenetic pace. Consumers and businesses are forced to keep upgrading software and sometimes to invest in new hardware in order to keep up.

What are the risks of such unlimited and unabated consumption? As has been noted, the high use of plastics and complex chemicals in the production of computer technologies is rarely considered an environmental problem by the computer industry, despite its need for nonrenewable resources such as silicon. Nor has the perpetual manufacturing and replacement of technologies in the workplace resulted in efficient production patterns that make good use of resources. While many corporations claim that they adopt one technology after another to increase productivity, studies indicate that greater productivity is rarely the result.[54] Productivity must be measured in terms of a society's total input and output. This means the negative consequences of technology-induced unemployment as well as social and environmental costs must be considered. When assessed in this way, it is clear that rapid technological progress has not delivered the efficiency it has promised.

Digital ideology has its eyes trained on the profits of a future that is built on the rhetoric of a friendly future of limitless possibility and the ultimate technological transcendence. Unfortunately, these myths are also myths built on the stark reality of silicon sweatshops and western technological restructuring — the burdens of which are most strongly felt by women.

Is This the Road Ahead, or the Road Back?[55]

This book began by reflecting on the very different ways that digital technology is affecting the lives of women who come from different socioeconomic groups and different countries. By examining *Wired* magazine, as an example of a particularly strident form of western digital discourse, I have tried to show how the current process of technological change is complemented by a discourse and ideology that reasserts power relations of inequality. There is an affinity between the digital ideology evident in magazines like *Wired* and the cultural narratives of modernity. The secular technological transcendence that David Noble argues arose as a result of western science's roots in the Latin Church seems evident in digital discourse as well. Through the construction of cyberspace as a form of digital escape, hypermacho man may well find the ultimate expression of otherwordly technological transcendence. This discourse is reinforced by a process of economic restructuring that seems to disadvantage the most vulnerable in our society and throughout the world. Our effort to crack the gender code of the emerging digital culture has revealed a close relationship between the interests of capital (particularly in the telecommunications and computer industries) and the new digital ideology that seeks to increase the use of these technologies in ways that reinforce sexism and racism. As a result, there is much more at stake in the drive to get online than simply the sale of computer equipment and subscriptions to Internet service providers. Far from providing a road to the future, digital discourse delivers a regressive ideology that exacerbates rather than alleviates women's inequality and social injustice.

What can be done? How can women respond to digital discourse and ideology?

Chapter Seven

BEYOND THE GENDER CODE: ASKING DIFFERENT QUESTIONS

> At *Wired* we keep our eyes trained on the future ... Our mantra is not *How has it been?* but *How will it be?*
>
> — Constance Hale, *Wired Style*

> Welcome to the Cybergrrl Webstation. A woman's place is online!
>
> — Cybergrrl Web site slogan
> <http://www.cybergrrl.com>

> There *are* alternatives to the capitalist utopia of total communication, suppressed class struggle, and ever-increasing profits and control that forgets rather than resolves the central problems of our society.
>
> — James Brook and Iain A. Boal,
> *Resisting the Virtual Life*

MOST OF US in the western world, who don't read *Wired* and aren't members of the elite digital generation, experience digital discourse through our role as consumers. We are familiar with a simpler form of digital discourse: advertisements. These ads do not tell us anything about the interests that are behind the digital revolution, who will benefit from this new digital age or by whose authority the "change leaders of our time" are creating our future. Our job, we are continually told, is to buy the products and get out of the way of progress. Clearly, then, the much-touted power of the consumer offers one arena for resistance. If we find digital discourse unappealing and potentially dangerous, we can decide not to buy equipment from companies that perpetuate myths of inequality and convince others not to buy magazines like *Wired* by publicly revealing its sexist and racist attitudes.

Yet many of us, as individuals, find we have to purchase computer equipment and software for our jobs or that our employers do so on our behalf. Our jobs often require that we continually update our skills in order to meet the demands of new technologies that are introduced into the workplace. Those of us who are not employed may be compelled to acquire training in the computer industry in order to find employment and may find ourselves vulnerable to a process of economic restructuring over which we have little control. As a result, for many of us, it seems impossible to participate in the decisions being made, let alone to opt out altogether. However, in our role as private consumers of computer and telecommunications advertising, we can learn to read between the lines to see where leading corporations are heading and how they are constructing our shared future. We can learn to reflect on these visions of the future and assess them, using them as a launching point from which to build our own hopes and aspirations. We can learn to identify the ideological assumptions that are embedded in popular discourse as a way of demystifying the speed and novelty of the information revolution. We can learn to be active consumers by demanding explanations for specific program functions and making our concerns known to corporate research and development departments. We can also demand to know more about how and under what conditions computer equipment is produced.

If we are really serious about engaging politically with digital discourse and ideology, we must make a determined effort to resist its assumptions. One of the ways that we can do that is by establishing a utilitarian relationship to technologies. That is, through our language and the way in which we use new technologies and think about them, we can put human priorities at the forefront. Computers are machines — masses of plastic, wire and silicon. They are our tools; we are not theirs. Just as there is more to the rhetoric of digital culture than just selling digital technology, there is more to being a citizen than just consuming. Or, in the words of Canadian social critic Irshad Manji, "Computers link up. Marketers network. Citizens are supposed to do much more."[1] Our first step to doing more is having the courage to name what is going on and validating our own feelings and experiences. The significance of this simple task cannot be overemphasized. Its importance was brought home to me by historian David Noble, who, in a CBC Radio broadcast described a telling experience he had with some of his young undergraduate students. Noble asked them to write down what they imagined their lives would be like twenty years from now. After the exercise, the students were asked whether they would mind if their statements were read out to the rest of the class. They expressed a great deal of anxiety about this possibility, only agreeing to the exercise once they were assured that the statements would be read without naming the writer. They needn't have demanded such discretion. The similarities between the statements were uncanny. And frightening. One after another, the students expressed the fear of being rendered "obsolete" by the computer revolution. One after another, they indicated a sense of a loss of control and agency in the construction of the future. In short, these undergraduates expressed an alarming degree of pessimism and insecurity. The cause? According to them, rapid technological change.[2]

Like Noble's students, many of us intuitively sense that something in the current process of technological change and widespread social restructuring does not jibe with the positive strains of corporate technotopic predictions. We feel uneasy. And our uneasiness is not an irrational fear of the unknown. We are uneasy not because we don't know what is really happening, but because we do know. We have seen an increase in economic insecurity and recognize a growing disparity between the rich and the poor. We have experienced the effects of unemployment and felt the stress of working harder and longer for less remuneration. We have come to expect less of all of our society's public institutions and feel our fates being determined by decision-makers to whom we have little or no access. Yet, while we see all these things, we feel powerless to effect change or even to articulate what is going on.

There are many reasons for this sense of powerlessness. The prevailing discourses of our times reinforce it. We are denied the roles of stakeholders in our future and of citizens in the present. Economists tell us high unemployment is both natural and functional. Governments tell us we are either in too much debt to deal with issues of social inequality or that we are subject to the larger realities of a globalized free market. Guide books about "surviving the information age" tell those of us who are feeling uncertain about new technologies that we need an "attitude adjustment."[3] As we have seen, the emerging discourse of digital technology also does little to empower individuals who are not members of the exclusive digital generation. Instead it exacerbates existing inequalities. We are swept up in the massive tide of this discourse, drifting farther and farther out in a sea of technological hallucinogens. The majority of us do not have the luxury of sober second thought and find ourselves struggling to adapt to changes beyond our control. What time is there for discourse analysis and democratic participation when it is all we can do to prevent ourselves from drowning? Yet, like the revelations of Noble's students, who discovered that they were not alone by revealing their views to one another, much of this isolation and confusion can be overcome by piecing together our collective experiences.

Fortunately, there are some very concrete strategies we can draw on to disturb the social conditions that perpetuate and enhance power relations of domination and subordination in a digital form. Before we explore such possibilities, we need to know exactly what we wish to achieve. We need to engage in some utopian fantasies of our own. What kind of information age would benefit the greatest number of people? How would new technologies fit into such a society? The problems associated with the proliferation of new digital technologies are too complex to be solved by simply providing universal access to information technologies. There is little point in having access to services that are dominated by consumer information and corporate interests, that create multiple levels of dependency and that facilitate a greater degree of worker surveillance and exploitation. We must step back much farther from the technologies themselves and determine what we are willing to sacrifice and what our requirements are from the information and communications sector. How important is job protection in the face of technological change? What kind of jobs should be saved? For whom? What about health, safety, the environment? What role do we see these technologies playing in the politics of democracy? How do we ensure that the current process of technological change does not perpetuate existing inequalities or create new ones?

These are difficult questions. Questions that I remain as uncertain of now as I did when I first felt uneasy about the way that digital discourse and ideology seems to be developing. The following discussion provides some suggestions about how we could develop a feminist politics of anticipation to understand the current digital revolution and politically engage with it. There are no definitive answers here, but there are some possibilities to consider as we make connections with other social groups, repoliticize the state and assess the role that digital technology should play in our strategies.

MAKING CONNECTIONS

As numerous political theorists and activists have noted, building coalitions across and within differences of class, race, gender, sexuality and so on is integral to any political strategy in the present era of fractured political resistance. The political landscape is increasingly characterized by numerous sites of rupture as high unemployment rates persist in the industrialized West and a mood of uncertainty and insecurity develop as a result of restructuring, globalization and technological change. There are many examples of social distress and anxiety — strikes, violent crime, cynicism about politicians and the democratic process, rising levels of racism and misogyny, to name just a few. Though ruptures in social peace and order do not provide positive indicators of the present/future, they do provide opportunities for creating powerful coalitions that bridge difference and wield significant strategic power.

Because technology has been largely addressed in the context of so-called objective science and inevitable progress, issues of technology and social change have often been neglected in social activist movements since the Luddite struggles of the British Industrial Revolution. At the same time, trade unions have become less and less effective in protecting workers from the impact of technological restructuring. Increased awareness of the myths of technology may help to ensure that such groups are able to more strongly articulate their members' needs in the future. Rather than responding to labour adjustments required by technological change in the workplace, trade unions may begin to play an enhanced role in determining what those changes will be.[4] The ability of social justice groups and trade unions to resist such inequality and exploitation may be strengthened by addressing issues of technological restructuring together. Recognizing and analyzing how technological change functions may help social movements to find new ways to mobilize and to effectively resist alienation and oppression.

It is clear that the future world envisaged by digital discourse affects different people differently, within and between race, gender and socioeconomic groups. The specific needs of the growing numbers of workers in untraditional jobs, like home telework, present new challenges that require leadership from the workers involved and a strong gender, class and race analysis of how these jobs fit into the new information economy. Existing unions must also recognize the profoundly gendered and racist nature of labour issues like job segregation, particularly in the high-technology sector. In combatting this problem, it is clear that simply re-educating women and racial minorities so that they are better able to survive in a globalized, competitive economy is not enough.[5] This strategy does little to address existing sexist and racist attitudes throughout society or to undermine the employment privilege currently enjoyed by upper-middle-class white men (and to a lesser extent, upper-middle-class white women). Focusing exclusively on the disadvantaged, without dismantling the privileges enjoyed by the advantaged, will not eliminate existing hierarchies between people or geographic regions. As the growing backlash to affirmative action programs (described as "reverse discrimination") indicates, we must expect that the status quo will be vigorously defended by those that benefit most from it.[6] Perhaps here is where re-education is really needed the most. Without government enforcement of policies that seek to remedy employment inequities and without the vigilant attention of social justice and labour groups, existing inequalities will continue in the digital age.

The diverse ways that digital technologies affect different people in different socioeconomic groups and regions of the world are complex. For example, for some women who work in microelectronics factories in the South, jobs with foreign multinational companies give them a greater degree of financial independence and personal choice than they have previously known and may provide a way out of the tyrannies of traditional societies. Nevertheless, these workers are also seen as an inexpensive, disposable and docile labour force to the companies that employ them. According to Swasti Mitter, it is important to recognize that "women in the third world welcome modernization, as long as they can have some say in the manner in which the technology which is affecting the quality of their working and family lives is adopted."[7] Strategies for change must have their origins in diverse locations and be co-ordinated at larger, international levels in order to help prevent conflicts, which will inevitably arise. Yet many of the problems we face, such as environmental degradation, are global in scope and therefore require global action. There are numerous places in which specific interests intersect and can be used as a common ground on which to build coalitions. It may be useful to contextualize the current process of technological change within the racist, classist and sexist history of western science and technology in order to help us recognize shared experiences. In any case, as we struggle to find common ground we must not reproduce dominant ideological assumptions of individual self-interest. We must avoid thinking of ourselves and others as stakeholders or clients with narrowly defined interests that we strive to secure in a hostile, zeroe-sum environment. A truly challenging and progressive movement demands that we think of ourselves as citizens who are concerned with the well-being of all other citizens and with the international community as a whole.

It is important to recognize that in this milieu of turbulence and anxiety, visions for a future technotopia become increasingly seductive in the mainstream. This has also occurred in academic institutions where connections to social and economic realities may be obscured. Alternative visions need to be articulated and disseminated to a wide audience. There is perhaps no stronger indication of a discontented society than the appeal of digital discourse, with its harsh edge and relentless desire for an easy technological solution. Ironically, it is here that one of the most compelling indications that social transformation is possible may lie. If there is genuine social change afoot in the present age of rapid technological development, it may be premised by the popularity of a disembodied, cybernetic virtual reality that takes the ultimate step of noncommittal through the modem. This regressive ideal may underline the degree to which existing constructions of white masculinity are no longer viable or desirable. In fact, it is these rigid constructions themselves that Lynne Segal suggests may be undermining men today.[8] Certainly these constructions are not healthy for any of us — men or women. With some of the most privileged members of society seeking escape, genuine calls for social change may not be far behind.

If we are at a point of crisis in western culture — a place of rupture and intensified constructions of gender and race — then there are many options for how we may proceed. Rather than following digital discourse down a road that intensifies inequality and perpetuates existing social, economic and environmental violence, we may be able to use this opportunity to reassess where we would really like to go from here. Feminist theory provides some important insights into the kind of connections we need to make between economic and social directions and the environment, between technological change and constructions of masculinity and femininity. If cyberspace is being constructed as the new Garden of Eden, this suggests that there is a real need to reassess our exploitative, alienated relationship to the natural world. How do we overcome the dominant instrumental view of nature in the West? Do we need to somehow "re-enchant" nature? How could we do this in a way that is authentic and resonates for men and women in the West? Can we imagine alternatives that emphasize rather than deny our existence as part of nature *and* culture?

Repoliticizing the State

If the basic premises of digital ideology become dominant, future political negotiation with the nation–state will undoubtedly become increasingly difficult for women. The (almost) complete removal of government from the regulation of information technology, coupled with the reliance on market forces and a discourse that perpetuates gender stereotypes, will continue to restrict women's access to the decision-making process in the field of technological change and the workplace. The myth of technological progress as inevitable and desirable will also continue to stifle debate about the social consequences of various technologies.

Fortunately, this is not a foregone conclusion. How can the state be used to redemocratize the current process of technological change?

Decisions about the new digital economy must be recognized to be political decisions, not simply technological ones. As we have seen, despite the much-discussed decline of the nation–state in the western world, national and supranational governmental bodies (like the International Monetary Fund, the World Bank, the Organization for Economic Cooperation and Development and the General Agreement on Tariffs and Trade, for example) do play a fundamental role in facilitating the development of the new digital economy. They must not, therefore, be let off the hook of public accountability. At the level of the nation–state, governments must be lobbied to ensure that local content legislation is strengthened to protect indigenous cultures, that privacy and security regulations are revamped to more adequately protect individual citizens and that environmental concerns are addressed in a progressive and timely fashion. The possibility of placing the information infrastructure under the control of an independent representative council that reflects diverse sectors of the global community, with an eye to redressing past inequities, should be seriously explored. Most importantly, the public must be alerted to the significance of the decisions currently being made about the future of all citizens. Trade agreements that dramatically affect the future of the world's economy, like the Multilateral Agreement of Investment (MAI), must not be negotiated under a veil of secrecy.

The existing powers of the state could be mobilized in a number of potentially helpful ways. As Heather Menzies suggests, progressive taxation measures that redirect the profits of transnational corporations for the benefit of people may be among the most practical and useful. A variety of possible systems could be implemented — for example, the imposition of a sort of "head tax" on computers to discourage rapid technological restructuring and the use of the so-called "Tobin tax" on foreign-exchange transactions or all digital information flows as a means of controlling the new economy. It would also be useful to conduct a computer census to keep track of how many computers are purchased during the same period as a given number of jobs are lost or social services are cut. This would certainly help to make the priorities of our present culture clear to the public. In addition, some kind of governmental (or third-party) regulation of the proliferation of information technology would be helpful in order to ensure that technologies that are put in place have positive effects on people, rather than simply eliminating or de-skilling their jobs.[9] Such methods would have to be implemented with the guidance of those individuals who are most vulnerable to the effects of corporate dominance — women, racial, ethnic and linguistic minorities, the poor, the disabled. If implemented sensitively, all of these strategies could be promising. However, we must be realistic not only about the difficulties of achieving the political will to employ such measures, but also about the possibility that, if legal restrictions like those envisaged by the proposed MAI are put in place, these measures may be beyond the jurisdiction of national governments.

It is also important to seriously assess the role of new information and communications technologies in the education system. Standards and implementation strategies for computer use in classrooms must be determined by the public and by professional educators, not by the computer industry. In addition, the gendered nature of technology and the exclusion of female and minority children from science and technology-related fields must be recognized in any curricula changes that are made. Creative computer games for boys and girls need to be developed that encourage children to explore technology without perpetuating gender stereotypes or violence. Science and technology-related fields themselves must become more open to debate and students in these fields must be required to place questions about science and technology within ethical and political frameworks. Compulsory, challenging courses in science studies and ethics for young engineers and computer scientists would help to create a sense of human responsibility in a world characterized by an unquestioned faith in science and technology.

MOBILIZING ONLINE?

We are barraged with images that depict the awesome strength and power of digital culture. Motorola Multimedia Group's "Power to the cable" advertisement provides a vivid example of how this power is portrayed (see Figure 5). The ad clearly links technological change with social revolution, suggesting that the use of Motorola's CableComm technology will bring forth a "bright new age" of interactivity — "the stuff that dreams are made of." The ad's textual message does little more than promise technotopia beyond the so-called digital revolution, but its graphic representations portray technology as a powerful force with enormous potential. The central image shows a male fist clutching a cable. Surrounding the fist is what looks like a series of orbiting planets or electrons. The image unifies science, technology and male power below the telling headline, "Power to the cable."[10] The smaller illustrations found in the corners of the ad support this message: a row of men hold cables like soldiers hold guns; a man victoriously thrusts a cable into the air; another man plugs a cable into a globe as an indication of his universal supremacy. With such powerful images of the masculine omnipotence of new technologies, it is hard to dispute the ad's final message: "The world is cable ready."

Advertisements like this one suggest that digital technologies are powerful tools of communication with important political consequences. While we may well object to the sexist, imperialistic elements of this message, it does not necessarily follow that the technologies themselves are inherently repressive. Many feminists and social activists have asked themselves whether the power of new information and communications technologies could be used as part of their own progressive political strategies. As we have seen, some groups have even taken on the task of making cyberspace a friendly place for women as their primary political goal. They have built "women-friendly" Web sites and written books to help women get online. These groups argue that it is more dangerous for women to risk being completely excluded from the world of digital technology than to work within an online environment that is often hostile to women. Women are told that they can meet people, make professional contacts and advance their careers by getting online. Sometimes, revolutionary discourse accompanies the vision of these groups. According to an article in *Women'space*, a Canadian women's Internet magazine, for example, women's drive to get online is part of a "cyber-revolution" — a movement that is compared with the bread riots of the French Revolution.[11] The same issue also features images of laptop-toting feminist activists, cartoons

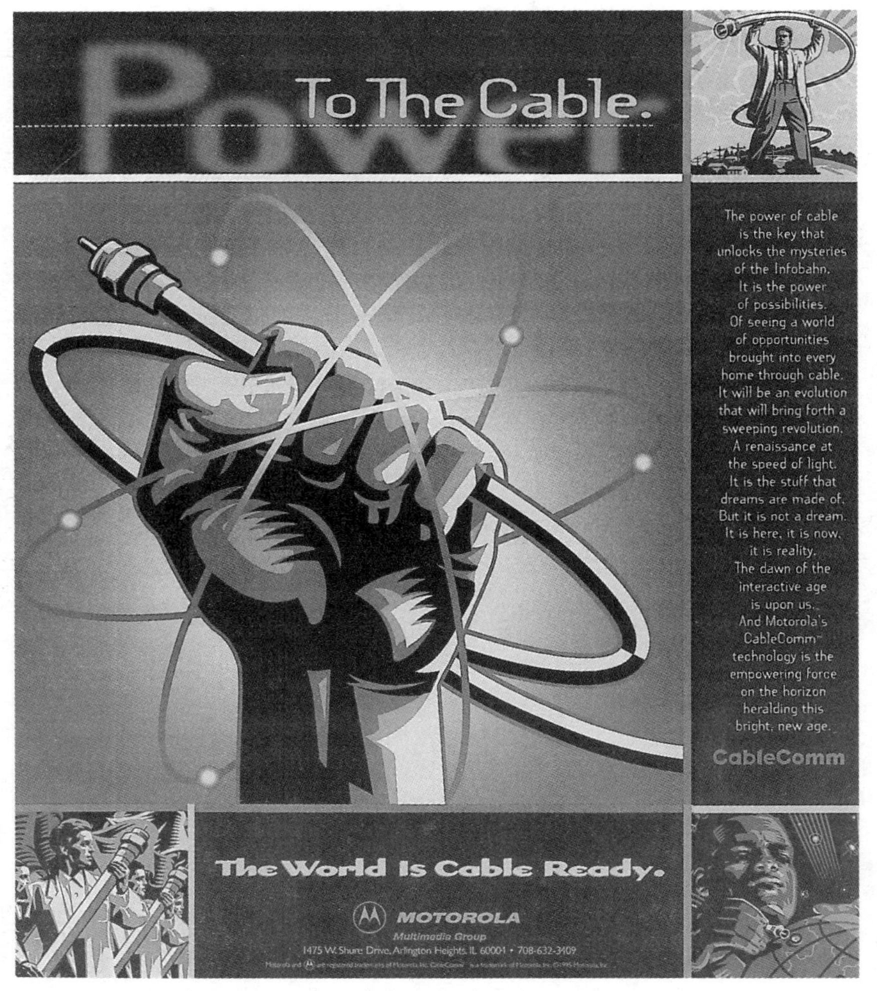

Figure 5: Digital ideology in a nutshell: technology + masculinity = power.

depicting women parachuting to earth with parachutes shaped like computer keyboards, and mouse cords wound into feminist symbols. Short blurbs describe sites of interest like the "Working Mom's Internet Refuge" and longer articles describe how to get online, train others and start connecting with other women's organizations. The images and articles are positive, attractive and evocative.

But there is also something vaguely disturbing about these "women-friendly" Web sites. Something that seems eerily reminiscent of the emerging digital discourse and ideology. Much of the same revolutionary language is used. Personal anecdotes tell readers about online successes. There is a tendency to glorify online activities. Flashy icons and short, superficial articles abound. For example, the Riotgrrl Web site features a cartoon of a voluptuous, scantily clad woman in stiletto heels, complete with a "bee hive" hair-do.[12] Meanwhile, visitors to the "Cybergrrl" Web site find images of the cartoon action heroine Cybergrrl, who tells women that "a woman's place is online!"[13] From this site women can link to the Clinique Cybergrrl column, and from there, click directly to Clinique's cosmetic Web site. Visitors are also invited to buy *Cybergrrl! A Woman's Guide to the World Wide Web*, a book written by Cybergrrl president Aliza Sherman.[14] The Web site ad for the book promises women that they will meet women online "discussing everything from parenting to starting a business to sewing to website design." In addition, visitors can connect to the Webgrrls "Biz Line" or hyperlink to www.careerbabe.com to network with new media businesswomen or pick-up some advice about online job searches. There is no message of political activism here beyond the urgent call, in the words of digital discourse, to "get wired."

Of course, not all online resources for women invoke traditional gender images and feature business-related themes. Some resources, including various list-servs and newsgroups, offer valuable information and a supportive, entertaining and helpful environment. Others, like The Feminist Majority Online and Virtual Sisterhood, play an important role in feminist networking. All are important in so far as their existence helps to disturb the cultural constructions of women as victims and as technologically incompetent. Yet lessons may be learned from certain images presented by advocates of the "cyber-revolution."

A recent cover illustration of *Women'space* magazine portrays a woman working on a computer and wearing a T-shirt bearing the words "Take Back the Night."[15] The computer screen displays a large feminist symbol. The image also contains a clock that reads 4:00 a.m., two empty coffee cups and a cat impatiently scratching on the window trying to get inside. A magazine on her desk is called *Insulating for Windows.* (See Figure 6.) While the illustration provides a graphic depiction of how women can use computer technology for feminist work, the T-shirt's slogan (which is the rallying cry for an annual night-time march at which Canadian feminists protest violence against women) tells another story that undermines the cartoon's positive message. Can women "Take Back the Night" without leaving their homes? Can organizing in cyberspace result in social change in the real world? Or does prolonged use of these technologies simply sap our energies, as the bags underneath the cartoon woman's eyes suggest? Are these Windows-based computer programs confining us to a more insular life in which we busily click our mouses as the world goes on

Figure 6: Selling digital technology as a tool for feminist activism. Is feminist cyber-revolution possible?

outside? Does the cat left outside represent the things we are neglecting as we pursue computer-mediated activism?

Unfortunately, some of the same companies that benefit from the discourse of *Wired* magazine will also benefit from the type of feminist discourse found in *Women'space* — Internet service providers, manufacturers of computers and software programs. As western feminists dial-up and log-on, we do so on equipment made by the same companies that exploit women's labour in developing countries, as is true of many of the products we consume. As we embrace the new digital information technologies, we must be aware that in doing so we risk reinforcing the same message conveyed by the Motorola advertisement (see Figure 5). By coming to rely on digital technologies in our own work, do we also give "Power to the cable"? Is it enough simply to replace the male fist with a female one?

New information and communications technologies *are* powerful tools that may be used effectively in a limited way by feminist organizations to communicate and exchange information. There is an element of truth in St. Jude's assertion that "girls *need* modems." However, just as domestic devices such as washing machines or dishwashers have been marketed to women in a way that reinforces the sexual division of labour,[16] so is digital technology being marketed to men in a way that reinforces a particular construction of hypermodern masculinity. The targeted end-user displays the characteristics of hypermodern machismo, is an ambitious entrepreneur and espouses libertarian views. In this context, it is important to raise the gender/technology question and to resist the sexist and racist elements of digital discourse. Doing so may help to destabilize the hegemonic acceptance of standardized technology as inevitable and neutral. By changing the perception of the end-user of digital technology as male through our engagement with the discourse and technology of digital culture, women may be able to begin the process of "degendering" the computer.[17] In so doing, women might just alter the ways in which digital technology is used and reinvent new information and communication technologies ourselves. Education, too, plays a key role in this process as we develop ways to subvert hegemonic messages of inequality and encourage female children to experiment with technologies in new and potentially liberatory ways and to invent new ones. In the meantime, women (and men) must confront online misogyny and racism anywhere they find it and take advantage of the (relative) safety afforded by technologically mediated communication to challenge locations that perpetuate inequality and minority exploitation.[18] Feminist engagement with digital technology through the modem must also occur across race, class and gender lines. Women must cast their net widely when (or if) they explore cyberspace. While the continued development of identity-specific Internet sites is important, such sites must not result in the ghettoization of women (and minorities) to specific corners of the "new frontier."

It is important not to simply adopt the structures and discourses of the new information and communications technologies. Susan Herring's research indicates that women already use computer-mediated communication differently than men, at least to some degree. While men are more likely to engage in aggressive debates and shun rules for discussion, women tend to use more polite and considerate modes of communicating.[19] By applying Eric von Hippel's concept of "users as innovators," it may be possible that women can influence the development of digital technology by using it in different ways.[20] To do so, we need to be more observant of and critically engaged in the technologies we use. For example, we can look at the reasons specific technologies (interfaces, logistics, design) cause us anxiety and assess the tasks we are trying to perform in order to find out what the social needs and outcomes of such technologies are. In the moments of frustration that frequently punctuate Internet experience lie important clues about what different individuals and collectivities require and feel when they use technologies. When we find ourselves feeling impatient with the thirty-second delay as bits travel along miles of cable to reach our computer screen, we need to recognize that such a response is a socially and technologically constructed one. We have come to value high-speed technology and expect it to deliver; this demand serves to perpetuate the hypermodern condition of immediate gratification that tends to disallow democratic debate and the careful development of social and economic policy. Perhaps we need to take the time to compare the privileges that high-speed digital culture affords us with the demands it places on our personal lives and on the social, economic and environmental reality. "Because we can" should never be a reason to integrate new programs and procedures. As feminists we must not depend on computer-mediated technologies as our only method of retrieving information and communicating with others. This dependency could result in our surrendering our ability to critically engage with digital technology and its discourse.

As we become more and more connected to the "wired world," we need to ask ourselves whether we are really being connected to more than enhanced possibilities for consumption. When I first started using e-mail, I was intrigued by its power and revelled in my new ability to keep in touch with friends and family who live elsewhere, while keeping my phone bill in check. However, it wasn't long before e-mail became yet another source of work-related demands on my limited leisure time. What I have saved in phone bills, I may have paid for in the loss of privacy and, to some degree, increased dependency. Cell phones may have a similar, though more intense, effect. Do we want to be available twenty-four hours a day? Is it fair for employers to assume they can contact an employee anytime, anywhere?[21] Is it possible to be too connected? Perhaps. Perhaps not. But we should be the ones to decide when, how and if we become wired.

As with any potential political strategy, there are no guarantees. For me, there remain many unanswered questions about the liberatory potential of new information and communications technologies and whether or not it is possible to summon the political will in our society to challenge corporate power and demand more than consumption. As we have seen, the social, cultural and economic context of these technologies is less than conducive to progressive social change. Yet while there may be no guarantees in political strategies, talking about their many advantages and disadvantages may help us avoid pitfalls and devise possible solutions to meet the barrage of hypermodernity with careful, considered strategies. As rapid technological development continues to shape western culture and society, feminists and social activists must carry on these discussions.

A FUTURE OF DIFFERENT QUESTIONS

No matter how it is packaged, technology is not a reified phenomenon. And technological change does not occur in a vacuum. Both are embedded in a complex system of power relations that for women has often meant the reassertion of beliefs and practices that have resulted in social and economic disadvantage. If *Wired* magazine is any indication of the direction that hegemonic western society is taking with respect to the meaning and context of the digital revolution, many of the most pressing issues of our hypermodern society — unemployment, widespread inequality, poverty — will be lost in a sea of technobabble and hypermacho libertarianism. We will continue the unself-conscious march into the future as the (un)willing pawns of technological developments that are being organized to serve the interests of a select few. Technological progress will continue to have the status of an unquestioned masculine deity and more and more of the world's population will be sacrificed to the needs of a superficial, empty form of progress.

Genuine social progress, it would seem, is not only notoriously slow but also rarely unidirectional. The resurgence of intense, overtly racist and sexist material in popular culture is not unique to *Wired* magazine. The present climate of rapid technological and social change has been accompanied by a reassertion of hegemonic power relations throughout western society. The age of backlash continues to dominate the portrayal of difference in popular culture. Even the well-respected Toronto *Globe and Mail Report on Business Magazine* has taken on elements of *Wired* style. Its premiere issue features a cover graphic that mimics hypermacho imagery beneath a caption telling us that Royal Bank's CEO "John Cleghorn Likes to Be On Top." Stories inside the issue feature catchy headlines like "Stealth Banker, "Barbie's at the Gate," "Moving Target," "Money Talks" and "Tech Know." The cover of the second issue mimics the gender stereotype of exceptional women who have achieved success based on their sexuality rather than on their business acumen. Two young businesswomen are photographed lying down and smiling up at the camera. According to the cover headline, their "aditude" is "turning the ad world upside-down." Editor Patricia Best even goes so far as to highlight the exceptionality of featuring women on the cover of a business magazine by dedicating her monthly message to explaining "How those two women got where they are." She need not have used such apologetic tones. The title of the article? "Hold the Bikini Babes."[22]

Wired is becoming less the extreme and more the norm. Why? As North Americans struggle to cope with increased levels of economic and political insecurity and rapid and relentless change, inequalities become more stark and the lines between "us" and "them" more clearly marked. It seems that we are living in a frightened society, and fear and insecurity never bring out the most generous human sentiments. The unequal relations of power discursively constructed by *Wired* are thus, to some degree, a result of the current socioeconomic context of uncertainty. *Wired* discourse is not just harvesting images of inequality from the past and repackaging them for our hypermodern times, it is constructing a new (post)human ideal for the future. It is a vision of digital escape: another in a series of masculine attempts to gain transcendence through technology. These are myths that are not going away. They are myths that we need to resist.

The emerging digital ideology is gaining strength in North American culture and has the potential to reach into every corner of the globe. It shows no signs of developing a sense of inclusive social responsibility and continues to perpetuate sexist and racist attitudes through its disturbing language and images. At the same time, the myths of digital ideology are being passed on to our children. These aspects of the emerging digital age are very alarming. Yet, for those of us who want to reintroduce justice and responsibility into our political and economic system, the most dangerous aspect of digital ideology is that it is closing the door on widespread debate and political participation. As the digital worldview becomes more common and colonizes more areas of the world with its masculine, techno-libertarian discourse, it stifles more alternative voices and paralyzes critique. If we allow it, this ideology will convince us that we are powerless to effect change and that if we oppose this corporate, technological form of progress we are Luddites, doomed to living in the past. If there has ever been a time when a wider discussion about the direction our future is taking is required, it is now. By developing and using a politics of anticipation, feminists and other social justice advocates can crack the gender code of digital culture and construct alternative visions for a society in which the goal is not to escape from the material world through some sort of elite technological salvation, but to actively participate within it.

Most of us are familiar with Microsoft's ubiquitous advertising slogan, "Where do you want to go today?" Perhaps now is the time to ask a profoundly different question: How do we make here a place where we want to stay — a place where we can all thrive?

❖

Notes

Introduction

1. Canadian College of Business and Computers advertisement, in partnership with IBM Canada. Toronto Transit Commission subway system, February 23, 1998.
2. This is not to suggest that there are no critical voices being raised to address the potential problems associated with digital technologies. However, such voices remain in the minority. Chapter 2 of this book discusses feminist approaches.
3. Liesbet van Zoonen, "Feminist Theory and Information Technology," *Media, Culture and Society* 14 (1992), 9–29.
4. It is clear that there is an artificial nature/nurture division being maintained here. As British academic Lynne Segal notes, when we attempt to distinguish between the social and the natural we frequently obscure the fact that "the one always already contains the other." *Slow Motion: Changing Masculinities Changing Men* (London: Virago Press, 1990), 64. What is important for this analysis is not to rigidly separate the two but to acknowledge the significance of cultural construction in the process of creating each and maintaining their separation.
5. Autumn Stanley, *Mothers and Daughters of Invention: Notes for a Revised History of Technology* (New Brunswick, NJ: Rutgers University Press, 1995), xxiii–xxvii.
6. In Stanley's words: "Anthropologists now generally agree that women invented agriculture," *Mothers and Daughters of Invention*, 1.
7. Ibid., 285. Italics in original.
8. A more thorough description of this process is provided in Chapter 1 of this book.
9. Stanley, *Mothers and Daughters of Invention*, 513.
10. Christine de Pisan, *The Book of the City of Ladies* (1405), trans. E. J. Richards (London: Picador, 1983).
11. Simone de Beauvoir, *The Second Sex* (1949), trans. H. M. Parshley (New York: Vintage, 1989), 64.
12. Stanley, *Mothers and Daughters of Invention*, xxxi–xxxiv.
13. Sadie Plant, *Zeros + Ones: Digital Women + The New Technoculture* (New York: Doubleday, 1997), 5–23.
14. Stanley, *Mothers and Daughters of Invention*, 438–442.
15. Judith A. Perrolle, *Computers and Social Change: Information, Property and Power* (Belmont, CA: Wadsworth, 1987), 67.
16. Plant, *Zeros + Ones*, 37.
17. Stanley, *Mothers and Daughters of Invention*, 433–512.
18. MUD stands for multi-user dungeon or domain, MOO for MUDs object-oriented. See glossary.

19. Rosie Cross, "Modem Grrl," *Wired*, February 1995, 119.
20. Ziauddin Sardar, "alt.civilizations.faq: Cyberspace as the Darker Side of the West," in Z. Sardar and J. R. Ravetz, eds., *Cyberfutures: Culture and Politics of the Information Superhighway* (New York: New York University Press, 1996), 24.
21. H. Jeannie Taylor, Cheris Kramarae, and Maureen Ebben, eds., *Women, Information Technology and Scholarship* (Urbana-Champaign, IL: The Center for Advanced Study, University of Illinois, 1993).
22. As quoted in Robert Chodos, Raue Murphy, and Eric Hamovitch, *Lost in Cyberspace? Canada and the Information Revolution* (Toronto: Lorimer, 1997), 75.
23. Carole A. Stabile, *Feminism and the Technological Fix* (Manchester: Manchester University Press, 1994), 4.
24. Essays like those contained in Lynn Cherny and Elizabeth Reba Weise, eds., *wired_women: Gender and New Realities in Cyberspace* (Seattle: Seal Press, 1996) are typical of feminist literature that reinforces the necessity of digital technology by providing personal accounts of how technically literate women can find alternative communities and entertainment online.
25. Melanie Stewart Millar, ed., "WomanTech," Special Issue, *WE International* 42/43 (Fall/Winter 1997).
26. Jay L. Lemke, *Textual Politics: Discourse and Social Dynamics* (London: Taylor and Francis, 1995), 2.

Chapter One

1. David F. Noble, *A World Without Women: The Christian Clerical Culture of Western Science* (New York: Alfred A. Knopf, 1992).
2. David F. Noble, *The Religion of Technology: The Divinity of Man and the Spirit of Invention* (New York: Alfred A. Knopf, 1997), 2–6.
3. In *The Religion of Technology*, Noble extends this connection further by explicitly arguing that technology has come to function as a religion of sorts in the modern western world.
4. Noble's recognition of the significance of science and technology to the rise of what he calls the "masculine millennium" is crucially important to understanding western history. In view of this, it is puzzling that Noble's latest work relegates the discussion of gender to an appendix (although a very lengthy one). The subordination of gender in the overall theoretical framework provided by Noble is odd because it would seem to strengthen the connection he makes between Christian and technological transcendence. Simone de Beauvoir's recognition of transcendence itself as primarily a masculine social endeavour would also seem an important, if neglected, one here. See Simone de Beauvoir, "Conclusion," *The Second Sex* (1949), trans. H. M. Parshley (New York: Vintage, 1989), 716–732.
5. Estimates about exactly how many women were murdered during this period range everywhere from several hundred thousand to nine million. It would seem likely

that a more accurate figure would lie somewhere between these. Autumn Stanley argues that even conservative estimates suggest that three million women were sacrificed during this period. Autumn Stanley, *Mothers and Daughters of Invention: Notes for a Revised History of Technology* (New Brunswick, NJ: Rutgers University Press, 1995), 98n11.

6. Carolyn Merchant, *The Death of Nature: Women, Economy and the Scientific Revolution* (London: Wilwood, 1980).
7. Michael Adas, *Machines as the Measure of Men: Science, Technology and Ideologies of Western Dominance* (Ithaca: Cornell University Press, 1989), 65–67.
8. Historians generally agree that the Industrial Revolution is more accurately described in terms of (at least) two separate revolutions, the first of which saw the development of the steam engine, spinning jenny and mechanization towards the end of the eighteenth century, and the second of which saw the development of electricity, the internal combustion engine and industrial chemicals roughly a hundred years later. For a brief overview, see Manuel Castells, *The Rise of the Network Society* (Malden, MA: Blackwell, 1996), 34–40.
9. For further discussion of the characteristics of modernity and postmodernity provided in this chapter, see David Harvey, *The Condition of Postmodernity* (Cambridge, MA: Blackwell, 1990); Derek Sayer, *Capitalism and Modernity: An Excursus on Marx and Weber* (London: Routledge, 1991); and Barbara L. Marshall, *Engendering Modernity: Feminism, Social Theory and Social Change* (Boston: Northeastern University Press, 1994).
10. Peter Wagner, *A Sociology of Modernity* (London: Routledge, 1993), 175.
11. See Iain A. Boal, "A Flow of Monsters: Luddism and Virtual Technologies," in J. Brook and I. A. Boal, eds., *Resisting the Virtual Life: The Culture and Politics of Information* (San Francisco: City Lights, 1995), 3–17. Today, it is common to call anyone opposed to technology and "progress" a Luddite. Some current critics of technology have described themselves as neo-Luddites and have sought to replace the universally negative views of the Luddite rebellions with more accurate depictions of these struggles as "pro-human," rather than simply "antitechnology."
12. While there remains a great deal of debate among feminists about exactly how capitalism and patriarchy interact, it is clear that the rise of the nuclear family in modern industrial society is of crucial importance to women's current social and economic location. For a good introductory discussion of this ongoing debate, see Valerie Bryson, *Feminist Political Theory* (New York: Paragon House, 1992).
13. Barbara L. Marshall, *Engendering Modernity: Feminism, Social Theory and Social Change* (Boston: Northeastern University Press, 1994), 9.
14. Harvey, *The Condition of Postmodernity*, 38.
15. Robert Chodos, Rae Murphy, and Eric Hamovitch, *Lost in Cyberspace? Canada and the Information Revolution* (Toronto: James Lorimer, 1997), 47.
16. Anne Balsamo, "Unwrapping the Postmodern: A Feminist Glance," *Journal of Communication Inquiry* 2, no. 1 (1987), 64–72.
17. Douglas Kellner, "The Postmodern Turn: Positions, Problems and Prospects," in

George Ritzer, ed., *Frontiers of Social Theory: The New Synthesis* (New York: Columbia University Press, 1990), 268.

18. David Noble, *Progress Without People: New Technology, Unemployment and the Message of Resistance* (Toronto: between the lines, 1995).
19. Chodos, Murphy, and Hamovitch, *Lost in Cyberspace?*, 111.
20. Ibid., 106–118.
21. David McIntosh, "Cyborgs in Denial: Technology and Identity in the 'Net," *Fuse* (Spring 1994),14–22.
22. Ziauddin Sardar, "alt.civilizations.faq: Cyberspace as the Darker Side of the West," in Ziauddin Sardar and Jerome R. Ravetz, eds., *Cyberfutures: Culture and Politics of the Information Superhighway* (New York: New York University Press, 1996), 16.
23. Third World Network, "Modern Science in Crisis: A Third World Response," in Sandra Harding, ed., *The Racial Economy of Science: Toward a Democratic Future* (Bloomington: Indiana University Press, 1993), 511.
24. Joni Seager, *The State of Women in the World Atlas* (London: Penguin, 1997), 69.
25. Swasti Mitter, "Information Technology and Working Women's Demands," in Swasti Mitter and Sheila Rowbotham, eds., *Women Encounter Technology: Changing Patterns of Employment in the Third World* (London: Routledge, 1995), 37.
26. Third World Network, "Modern Science in Crisis," 499. This view is also expressed by popular economist and best selling American author William Greider who argues that "women are the most exploited class in the global industrial system." *One World, Ready or Not: The Manic Logic of Global Capitalism* (New York: Simon and Schuster, 1997), 469.
27. Mitter, "Information Technology and Working Women's Demands," 29.
28. Third World Network, "Modern Science in Crisis," 510.
29. For an excellent discussion of biotechnology and the erosion of the world's biodiversity, see Vandana Shiva, *Biopiracy: The Plunder of Nature and Knowledge* (Toronto: between the lines, 1997).
30. Jeremy Rifkin, *The Biotech Century: Harnessing the Gene and Remaking the World* (New York: Jeremy P. Tarcher/Putnam, 1998), 175, 192.
31. Castells, *The Rise of the Network Society*, 13. Castells even goes so far as to argue that "without new information technology global capitalism would have been a much-limited reality," 19.
32. Linda McQuaig, *The Cult of Impotence: Selling the Myth of Powerlessness in the Global Economy* (Toronto: Viking Penguin, 1998).
33. Tony Clarke and Maude Barlow, *MAI: The Multilateral Agreement on Investment and the Threat to Canadian Sovereignty* (Toronto: Stoddart, 1997).
34. To add further irony to the story of the recently suspended MAI negotiations, the Internet itself may have played an important role in scuttling the deal. The globalized computer network was used by the opposition to effectively communicate new developments and mobilize public resistance to the treaty in a timely fashion. It remains debatable whether or not such opposition could have been mobilized

without the use of electronic communications.

35. Heather Menzies argues that every sector of the economy, "from mining to manufacturing to the giving and receiving of services," is being affected by the current process of technological change. *Whose Brave New World? The Information Highway and the New Economy* (Toronto: between the lines, 1996), xvi, 7–8.

36. Ray Marshall, "The Global Jobs Crisis," *Foreign Policy* (Fall 1995), 50.

37. Stanley Aronowitz and William DiFazio, *The Jobless Future: High-Tech and the Dogma of Work* (Minneapolis: University of Minnesota, 1994).

38. Donald Tomaskovic-Devey, *Gender and Racial Inequality at Work: The Sources and Consequences of Job Segregation* (Ithaca, NY: ILR Press, 1993), 7.

39. Barbara Reskin and Patricia A. Roos, *Job Queues, Gender Queues: Explaining Women's Inroads into Male Occupations* (Philadelphia: Temple University Press, 1990).

40. Tomaskovic-Devey, *Gender and Racial Inequality at Work*, 11.

41. Ibid., 12, 16, 137.

42. Sara Diamond, "Taylor's Way: Women, Cultures and Technology," in Jennifer Terry and Melodie Calvert, eds., *Processed Lives: Gender and Technology in Everyday Life* (New York: Routledge, 1997), 88.

43. Seager, *The State of Women in the World Atlas*, 70.

44. For detailed accounts of the impact of computer technology on women's employment in western industrialized countries, see: Eileen Green et al., *Gendered by Design?* (London: Taylor and Francis, 1993); H. Hartmann, ed., *Computer Chips and Paper Clips* (Washington: National Research Council, 1986); Ellen Lupton, *Mechanical Brides: Women and Machines from Home to Office* (New York: Cooper-Hewitt, 1993); J. Morgall, "Typing Our Way to Freedom," *Feminist Review* 9 (1981), 87–103; Barbara Drygulski Wright, ed., *Women, Work and Technology: Transformations* (Ann Arbor: University of Michigan Press, 1987).

45. Fiona Wilson "Women, Office Technology and Equal Opportunities: The Role of Trade Unions," in M. J. Davidson and C. L. Cooper, eds., *Women and Information Technology* (New York: John Wiley and Sons, 1987), 243.

46. Barbara S. Burnell, *Technological Change and Women's Experience: Alternative Methodological Perspectives* (Westport, CT: Bergin and Gavey, 1993), 168–169.

47. Wilson, "Women, Office Technology and Equal Opportunities: The Role of Trade Unions," 243.

48. Burnell, *Technological Change and Women's Experience*, 169.

49. Tomaskovic-Devey, *Gender and Racial Inequality at Work*, 3.

50. Yanick St. Jean and Joe R. Feagin, "Black Women, Sexism and Racism," in Carol Rambo Ronai, Barbara A. Zsembik, and Joe R. Feagin, eds., *Everyday Sexism in the Third Millennium* (London: Routledge, 1997), 158. There is also a lack of research on how different ethnic groups have experienced technological change. As a result, the discussion provided here is unsatisfactory and deals almost exclusively with existing research on African-American women's experience.

51. Venus Green, "Race and Technology: African American Women in the Bell System, 1945–1980," *Technology and Culture* 36, no. 2 (1995), S601–S643; Evelyn Nakano Glenn and Charles M. Tolbert II, "Technology and Emerging Patterns of Stratification for Women of Color: Race and Gender Segregation in Computer Occupations," in Wright, ed., *Women, Work and Technology.*
52. Green, "Race and Technology," S109.
53. Tomaskovic-Devey, *Gender and Racial Inequality at Work*, 137.
54. The legacy of slavery plays an important role in determining how African-American women are treated on the job. See St. Jean and Feagin, "Black Women, Sexism and Racism," 157. According to Tomaskovic-Devey, African-American women are supervised more closely than white women on the job. He suggests that this may be the result of white male assumptions that white women can be treated like wives, who are trusted to carry out their tasks both at work and at home. (*Gender and Racial Inequality at Work*, 14.) It is important to recognize, however, that women in general have been found to be supervised more on the job than men. Menzies, *Whose Brave New World?*, 118.
55. St. Jean and Feagin, "Black Women, Sexism and Racism," 159.
56. Despite the fact that there is no evidence to support it, Tomaskovic-Devey's research found that white male managers still make the assumption that women and African-Americans are more costly to train and less productive than white male workers. When training costs are perceived as high — a situation that is particularly true in the computer industry — it is clear that "white males will tend to be preferred" for such jobs. *Gender and Racial Inequality at Work*, 9.
57. St. Jean and Feagin, "Black Women, Sexism and Racism," 178.
58. Wendy Dawson, *When She Goes to Work She Stays At Home* (Canberra: Australian Government Documents, 1989); Menzies, *Whose Brave New World?*, 34-35.
59. I refer here to women with responsibilities to care for children, the elderly or the disabled, as well as those women who have mobility problems due to their age, disability or remote location.
60. It is important to recognize that not all western women have the same relationship to the traditional ideology of domesticity. For many women of colour and poor women, staying home to raise children has simply never been an option. Yet the ideology of domesticity remains intimately linked to dominant beliefs about femininity.
61. Pam Evans, "Women, Homeworkers, and Information Technology," in M. J. Davidson and C. L. Cooper, eds., *Women and Information Technology* (New York: John Wiley and Sons, 1987), 228.
62. Ibid., 226–229.
63. Raija Kalemo and Anneli Leppanen, "Video Display Units — Psychosocial Factors in Health," in M. H. Davidson and C. L. Cooper, eds., *Women and Information Technology* (New York: John Wiley and Sons, 1987), 193–227.
64. Chodos, Murphy, and Hamovitch, *Lost in Cyberspace?*, 123.

65. Vernon L. Mogensen, *Office Politics: Computers, Labor and the Fight for Safety and Health* (New Brunswick, NJ: Rutgers University Press, 1996), 143, 144, 145.

66. Mary Sue Henifin, "The Particular Problems of Video Display Terminals," in Wendy Chavkin, ed., *Double Exposure* (New York: Monthly Review Press, 1984), 69–80; Mogensen, *Office Politics*, 7.

67. Menzies, *Whose Brave New World?*, 32.

68. Discrimination against women occurs within computer training seminars, computer science courses, and other science-related contexts. See Martha A. Bartter, "Science, Science Fiction and Women: A Language of (Tacit) Exclusion," *Etc.* 49 (Winter 1992), 407–419; P. Edwards, "The Army and the Microworld: Computers and the Politics of Gender Identity," *Signs* (Autumn 1990), 102–127; Sally Hacker, *Doing It the Hard Way: Investigations of Gender and Technology* (Boston: Unwin Hyman, 1990); Tov Hapnes and Bente Rasmussen, "Excluding Women from the Technologies of the Future? A Case Study of the Culture of Computer Science," *Futures* 23 (December 1991), 107–111; Fergus Murray, "A Separate Reality: Science, Technology and Masculinity," in E. Green et al., eds., *Gendered by Design?* (London: Taylor and Francis, 1993); Margie Wylie, "No Place for Women," *Digital Media* 4, no. 8 (January 1995), 3–14; Deborah Sturm and Marsha Moroh, "Encouraging Enrollment and Retention of Women in Computer Science Classes," *Output* (January 1995), 28–29.

69. In fact, *Wired* completely dismisses claims of unequal access. In the January 1998 issue, *Wired* reproduced the graphics from the same Compaq ad, replacing the original text with the following: "The real social threat is not that everyone won't be connected — but that no one will be able to disconnect." "Change is Good: The State of the Planet 1998," *Wired*, January 1998, 198–199.

70. Obviously this is not a simple process. Individuals do not just reproduce images of gender found in hegemonic culture. However, over time, these images do inscribe themselves on our psyches — we are not immune to their effects, even when we think that we are subverting or ignoring them.

71. White heterosexual women, like white heterosexual men, remain the normative models for dominant constructions of masculinity and femininity in western culture.

72. Margaret Lowe Benston, "A New Technology But the Same Old Story," *Canadian Woman Studies/Les cahiers de la femme* 13, no. 2 (Winter 1993), 68–81.

73. Seager, *The State of Women in the World Atlas*, 60, 66, 78.

74. Janine Brodie's *Politics on the Margins: Restructuring and the Canadian Women's Movement* (Halifax: Fernwood Publishing, 1995) provides an excellent overview of how restructuring has affected the lives of women and how the process itself is an undeniably gendered one.

75. Susan Faludi, *Backlash: The Undeclared War on American Women* (New York: Doubleday, 1991).

76. Canadian political scientist Sylvia Bashevkin extends Faludi's project to the conservative politics of the Thatcher years in Britain, Reagan's United States and the Mulroney era in Canada. Unfortunately, she notes that the replacement of these

figures with the more "moderate" regimes of Tony Blair, Bill Clinton and Jean Chrétien does not appear to have returned the debate to a more progressive discourse. Women continue to be largely "on the defensive." *Women on the Defensive: Living Through Conservative Times* (Toronto: University of Toronto Press, 1998).

77. Roberta Furger, *Does Jane Compute? Preserving Our Daughters' Place in the Cyber Revolution* (New York: Warner Books, 1998), 154.

78. The gendered nature of computer and telecommunications industry advertising with respect to the Information Highway is pronounced. Several examples have been highlighted in this analysis. See also, Jane Caputi, "Seeing Elephants: The Myths of Phallotechnology," *Feminist Studies* 14 (Fall 1988), 487–525; and Melanie Stewart Millar, "*Wired* Words/ *Wired* Worlds: The Construction of Race, Class, and Gender in Digital Discourse," paper presented at the Race, Gender and the Construction of Canada Conference, University of British Columbia, October 21, 1995.

79. Carol J. Adams, "'This Is Not Our Fathers' Pornography': Sex, Lies, and Computers," in Charles Ess, ed., *Philosophical Perspectives on Computer Mediated Communication* (Albany, NY: State of New York University Press, 1996), 154–156.

80. While the "blockbuster hit of 1996" for boys was Duke Nukem 3D (see Figure 5), a violent fighting game, Barbie Fashion Designer was the best-seller in the genre of girls software in the same year. Roberta Furger, *Does Jane Compute?*, 56.

81. Purple Moon advertising slogan.

82. Rebecca L. Eisenberg's, "Girl Games: Adventures in Lip Gloss," *Ms. Magazine*, January/February 1998, 84–87; Roberta Furger, *Does Jane Compute?*, 46–48.

83. Hopefully, as Furger suggests, gender neutral games will become more readily available as game companies realize their potential for universal appeal. *Does Jane Compute?*, 52.

84. Debbie Stoller, "Grrls and their Games," *Shift*, June 1998, 94; Clive Thompson, "Girl Gamers Battle Boy Bias," *The Globe and Mail*, 6 June 1998, C1.

85. See Carol Rambo Ronai, Barbara A. Zsembik, and Joe R. Feagin, eds., *Everyday Sexism in the Third Millennium* (London: Routledge, 1997). Readers may find this collection of articles interesting. It provides an overview of how sexism is in fact an enduring phenomenon — despite conservative rhetoric that suggests that such problems are "solved."

86. Barbara Ehrenreich, *The Hearts of Men: American Dreams and the Flight from Commitment* (Garden City, NY: Anchor Books, 1983).

87. While Lynne Segal suggests that images of masculinity have become more closely related to the family as "fatherhood" has become more significant, she also suggests that this has not resulted in a shift of power. She argues that while men may be taking parenting more seriously in the 1990s than in the 1950s (in terms of their direct involvement), they are far from genuinely sharing responsibilities. *Slow Motion: Changing Masculinities Changing Men* (London: Virago Press, 1990), 37.

88. Arthur Kroker and Michael A. Weinstein, *Data Trash: The Theory of the Virtual Class* (New York: St. Martin's Press, 1994), 41.

Chapter Two

1. Heather Menzies, *Whose Brave New World? The Information Highway and the New Economy* (Toronto: between the lines, 1996); David F. Noble, *The Religion of Technology: The Divinity of Man and the Spirit of the Inventor* (New York: Knopf, 1997); David F. Noble, *Progress Without People: New Technology, Unemployment and the Message of Resistance* (Toronto: between the lines, 1995); Neil Postman, *Technopoly: The Surrender of Culture to Technology* (New York: Vintage Books, 1992); Langdon Winner, *The Whale and the Reactor: A Search for Limits in an Age of High Technology* (Chicago: University of Chicago Press, 1986).
2. Derrick de Kerckhove, *The Skin of Culture: Investigating the New Electronic Reality* (Toronto: Somerville House, 1995); Nicholas Negroponte, *Being Digital* (New York: Knopf, 1995).
3. de Kerckhove, *The Skin of Culture,* 140.
4. Negroponte, *Being Digital,* 231.
5. Bell Canada advertising slogan.
6. Estimates of the exact number of women that are online are remarkably diverse. Ziauddin Sardar argues that "less than one percent of the people online are women," while Taylor et al. argue that between 10 and 15 percent are women. In any case, the vast majority of scholars and researchers agree that men outnumber women. The cyberfeminist Sadie Plant is a notable exception. She suggests that 50 percent of Net users are female. However, as she herself notes, it is difficult to determine the accuracy of demographic information in cyberspace. See Ziauddin Sardar, "alt.civilizations.faq: Cyberspace as the Darker Side of the West," in Z. Sardar and J. Ravetz, eds., *Cyberfutures: Culture and Politics of the Information Superhighway* (New York: New York University Press, 1996), 24; H. J. Taylor, C. Kramarae and M. Ebben, *Women, Information Technology and Scholarship* (Urbana-Champaign, IL: The Center of Advanced Study, University of Illinois, 1993); Sadie Plant, *Zeros + Ones: Digital Women + the New Technoculture* (New York: Doubleday, 1997), 112.
7. Margie Wylie, "No Place for Women," *Digital Media* 4, no. 8 (January 1995), 3.
8. Liesbet van Zoonen, "Feminist Theory and Information Technology," *Media, Culture and Society* 14 (1992), 11.
9. The ARPANET project was a U.S. military project that began in 1969. It evolved into a primarily research-related link between academic institutions before becoming accessible to individual users.
10. I have not attempted to discuss all the many different and nuanced approaches that feminists have taken to new technologies. Rather, I have provided an overview here of those that are dominant, on-the-rise or of particular concern to this book.
11. Nancy Kaplan and Eva Farrell, "Weavers of Webs: A Portrait of Young Women on the Net," *Arachnet Electronic Journal of Virtual Culture* 2, no. 3 (July 1994), as quoted in Dale Spender, *Nattering on the Net: Women, Power and Cyberspace* (Toronto: Garamond, 1996), 233–234.

12. Peregrine Wood, "Changing the World Electronically," *WE International* 42/43 (Fall/Winter 1997), 35–39.

13. Spender, *Nattering on the Net*, 249.

14. Ibid., 69.

15. Susan Myburgh, "Cyberspace: A New Environment for Women," *WE International.* 42/43 (Fall/Winter 1997), 20–24.

16. Ibid., 22.

17. Plant, *Zeros + Ones*, 38.

18. Plant outlines her cyberfeminism most clearly in "Beyond the Screens: Films, Cyberpunk and Cyberfeminism," *Variant* 14 (1993), 13–17.

19. Donna Haraway, "A Manifesto for Cyborgs: Science, Technology and Socialist Feminists in the 1980s," *Socialist Review* 80, no. 2 (1985), 65–108. Haraway's most recent offering *Modest_Witness@Second_Millennium.FemaleMan©_Meets_OncoMouse™: Feminism and Technoscience* (New York: Routledge, 1997), is probably her least accessible work and continues her theoretical explorations of cyborgs, biotechnology and cyberspace.

20. Plant, "Beyond the Screens: Film, Cyberpunk and Cyberfeminism," 13.

21. See Janet Biehl, *Rethinking Ecofeminist Politics* (Boston: South End Press, 1991).

22 Sherry Turkle, *Life on the Screen: Identity in the Age of the Internet* (New York: Simon and Schuster, 1995), 263.

23. Ibid.

24. Claudia Springer, *Electronic Eros: Bodies and Desire in the Postindustrial Age* (Austin: University of Texas Press, 1996), 11.

25. For discussion of women's lower levels of online participation see Susan Herring, "Gender and Democracy in Computer Mediated Linguistic Communication," *Electronic Journal of Communication* 3, no. 2 (1993).

26. See Rodney Peterson, "Harassment by Electronic Mail," *Synthesis: Law and Policy in Higher Education* (Winter/Spring 1994), 402–403, 416; Kimberly J. Cook and Phoebe M. Stambaugh, "Tuna Memos and Pissing Contests: Doing Gender and Male Dominance on the Internet," in C. Rambo Ronai, B. A. Zsembik, and J. R. Feagin, eds., *Everyday Sexism in the Third Millennium* (New York: Routledge 1997), 63–83; Cheris Kramarae and H. Jeannie Taylor, "Women and Men on Electronic Networks: A Conversation or a Monologue?," in H. Jeanie Taylor, Cheris Kramarae, and Maureen Ebben, eds., *Women, Information Technology and Scholarship* (Champaign-Urbana, IL: The Center for Advanced Study, University of Illinois, 1993), 52–61; Pamela J. Cushing, "Gendered Conversational Rituals on the Internet: An Effective Voice is Based On More Than Simply What One is Saying," *Anthropologica* 38 (1996), 47–80.

27. See Larissa Silver, "Same Message, Different Medium," *WE International* 42/43 (Fall/Winter 1997); Susan Herring, Deborah A. Johnson, and Tamra Di Benedetto, "This Discussion Has Gone Too Far!: Male Resistance to Female Participation on the Internet," in Kira Hall and Mary Bulcholtz, eds., *Gender Articu-*

lated: Language and the Socially Constructed Self (New York: Routledge, 1995); Cheris Kramarae and Jana Kramer, "Net Gains, Net Losses," *Women's Review of Books* (February 1995), 33–35; Susan Herring, "Politeness in Computer Culture: Why Women Thank and Men Flame," in Mary Bulcholtz, Anita Liang, and Lauren Sutton, eds., *Communicating in, through, and across Cultures: Proceedings of the Third Berkeley Women and Language Conference* (Berkeley, CA: Berkeley Women and Language Group, 1994), 278–294; Anne Balsamo, "Feminism for the Incurably Informed," in Mark Dery, ed., *Flame Wars: The Discourse of Cyberculture* (Durham, NC: Duke University Press, 1994), 125–156.

28. Plant, *Zeros + Ones*, 41–42.
29. Joni Seager, *The State of Women in the World Atlas* (London: Penguin, 1997), 74.
30. Carol J. Adams, "'This Is Not Our Fathers' Pornography': Sex, Lies and Computers," in Charles Ess, ed., *Philosophical Perspectives on Computer-Mediated Communication* (Albany, NY: State University of New York Press, 1996), 157.
31. Stimulated by the insights of the women's movement and the crisis of modernity, the study of women's relationship to language emerged alongside discourse analysis. Since the 1970s, the field has grown substantially. Now, discourse analysis is pursued in numerous diverse academic disciplines. In many ways, the development of the argument that gender is discursively constructed is the logical extension of one of feminism's most significant insights, that while sex is rooted in biology, gender is a socially constructed phenomenon.
32. Norman Fairclough, *Language and Power* (London: Longman, 1989), 3.
33. Menzies, *Whose Brave New World?*, 122.
34. Anne Balsamo, *Technologies of the Gendered Body: Reading Cyborg Women* (Durham, NC: Duke University Press, 1996).
35. Springer, *Electronic Eros*, 10.
36. Ibid., 161.
37. Balsamo, *Technologies of the Gendered Body*, 16.
38. Carole A. Stabile, *Feminism and the Technologial Fix* (Manchester: Manchester University Press, 1994), 156, 157, 158. For this reason, Swasti Mitter and Sheila Rowbotham's collection of essays exploring Third World women's experience of technological change is extremely valuable: *Women Encounter Technology: Changing Patterns of Employment in the Third World* (London: Routledge, 1995). There is a real need for further research in this area and for analyses that explore the linkages between diverse regional experiences.
39. Stabile, *Feminism and the Technologial Fix*, 157.
40. The dangers of an exclusive focus on class were starkly illustrated recently at the conference "Challenging Corporate Rule." The workshop on "The Global Pillage: Telecommunications and the Information Highway" recognized class but completely ignored the fact that "corporate rule" is experienced in many ways and is influenced not only by class location but also by such variables as race and gender. This omission was particularly surprising in view of the fact that much of the workshop dealt with the struggles of Canadian telephone industry workers — a

sector of the economy in which women are well represented and which has been the focus of a significant amount of feminist research. "Challenging Corporate Rule: A Global Teach-In Conference," sponsored by the Council of Canadians, University of Toronto, November 7–9, 1997.

41. While it is impossible to say for sure whether digital discourse will become hegemonic in the future, it is becoming increasingly influential. This alone justifies the need for critical feminist analysis of its structures and meanings.
42. Fiona Wilson, "Language, Technology, Gender and Power," *Human Relations* 45 (September 1992), 886.
43. My own experience testifies to this. The unfamiliar, technical nature of *Wired* discourse frequently forced me to ask for assistance from more technologically savvy friends and family.
44. During my study of *Wired* magazine, I surveyed every issue of the magazine since its inception in January 1993.
45. See, for example, Teun van Dijk, "Principles of Critical Discourse Analysis," *Discourse and Society* 4, no. 2 (1993), 249–283, and, van Dijk, ed., *Discourse Studies: A Multidisciplinary Introduction* (London: Sage, 1997); Norman Fairclough, *Language and Power* (London: Longman, 1989); Roger Fowler, "Power," in Teun van Dijk, ed., *Handbook of Discourse Analysis*, vol. 4 (London: Academic Press, 1985), 61–82; Teun van Dijk, *Linguistic Criticism* (Oxford: Oxford University Press, 1990); Stuart Hall, "Encoding/Decoding," in S. Hall, D. Hobson, A. Lowe, and P. Willis, eds., *Culture, Media, Language* (London: Hutchinson, 1980); Deborah Cameron, ed., *The Feminist Critique of Language: A Reader* (London: Routledge, 1990); Julia Penelope, *Speaking Freely: Unlearning the Lies of the Fathers Tongues* (New York: Pergamon Press, 1990); Liesbet van Zoonen, *Feminist Media Studies* (London: Sage, 1994); Susan Ehrlich and Ruth King, "Feminist Meanings and the (De)politicization of the Lexicon," *Language in Society* 23 (1994), 59–76.
46. It is assumed for the sake of this analysis that individual authors, aware of this preexisting symbolic order, use it to help create new understandings, together forming an identifiable digital discourse. This discourse is then taken up and interpreted by readers or subjects of the discourse, who interpret its meanings based on shared symbolic associations and cultural motifs. While this process is most emphatically not an uncomplicated one and cannot be fully realized without the integration of reader–response survey results, it is possible to make limited assumptions about the relative success of particular discourses based on their rates of growth and the degree to which the discourse seeps into the mainstream. In the case of *Wired* magazine, we will see that both of these are very high.
47. Duncan Davies, Diana and Robin Bathurst, *The Telling Image: The Changing Balance Between Pictures and Words in a Technological Age* (New York: Clarendon Press, 1990), ix.
48. The necessity of studying text and graphics together is manifest when we consider the construction of meaning in print and television advertisements. The interplay of words and images in advertisements is readily apparent — but how is this relationship to be understood? The work of semiotics, the study of signs, provides im-

portant clues about this relationship. For example, visual semiotics allow us to literally map how visual space is given meaning in western cultural discourse. While this is not the place for a discussion of discourse semiotics, an excellent introductory explication is provided by Gunther Kress, Regina Leite-Garcia, and Theo van Leeuwen, "Discourse Semiotics," in van Dijk, ed., *Discourse Studies: A Multidisciplinary Introduction*, 257–289.

49. Constance Hale, ed., *Wired Style: Principles of English Usage in the Digital Age* (San Franscisco: HardWired Books, 1996), 66.

Chapter Three

1. Heather Menzies, *Whose Brave New World? The Information Highway and the New Economy* (Toronto: between the lines, 1996), 47.
2. Paulina Borsook, "The Memoirs of a Token: An Aging Berkeley Feminist Looks at *Wired*," in Lynn Cherny and Elizabeth Reba Weise, eds., *wired_women: Gender and New Realities in Cyberspace* (Seattle: Seal Press, 1996), 26.
3. Kristy O'Rell, *Wired* Customer Service Representative, e-mail to author, 29 August 1995.
4. Interestingly, this demographic strongly resembles the typical online subscriber described by a recent Prodigy survey and shares general characteristics with the *Playboy* demographic. The Prodigy survey found most subscribers were male, professionals, married with children and earning an above average income. Michael Newbarth, "Virtual El Dorado," *Internet World*, June 1995, 8. *Playboy*, the men's pornographic magazine, has a demographic that includes men aged 18 to 35, who are well-educated and earning a higher than average income. *Playboy* demographic as described by Ziauddin Sardar, "alt.civilizations.faq: Cyberspace as the Darker Side of the West," in Z. Sardar and J. R. Ravetz, eds., *Cyberfutures: Culture and Politics of the Information Superhighway* (New York: New York University Press, 1996), 25.
5. See "*Wired* Marketing Information," 1997 Advertising Kit, 1996 Intelliquest V3.0/Businesss Study, "*Wired* to the Net," Beta Research, "1997 Subscriber Study," "*Wired* Readers Take Action," and "A Passion for the *Wired* Point of View." While these statistics are not beyond suspicion, given that they are published by *Wired* itself and clearly serve its interests, they remain significant. Even if greatly exaggerated, they are supported by *Wired*'s circulation success as indicated by figures provided by the Audit Bureau of Circulations.
6. *Mondo 2000*, September 1990, cover.
7. As quoted in Herbert I. Schiller, "The Global Information Highway: Project for an Ungovernable World," in James Brook and Iain A. Boal, eds., *Resisting the Virtual Life: The Culture and Politics of Information* (San Francisco: City Lights, 1995), 17.
8. Howard Besser, "From Internet to Information Superhighway," in Brook and Boal, eds., *Resisting the Virtual Life*, 59–71.
9. See *Wired* 1997 Advertising Kit.

10. Audit Bureau of Circulations, "*Wired* Marketing Information," as quoted in *Get Wired* Advertising Kit (San Francisco: Wired Ventures, 1995).

11. See *Wired* 1997 Advertising Kit.

12. Audit Bureau of Circulations, *FAS-FAX* (Shaumberg, IL: Audit Bureau of Circulations, June 1997).

13. Total paid subscriptions for the period ending June 1997: *PC World*, 1,143,710; *PC/Computing*, 1,006,285; *GQ*, 688,752; *Business Week* (N.A.), 915,149; *Fortune* (N.A.) 762,701; *Rolling Stone*, 1,256,915; *Wired*, 347, 465. Audit Bureau of Circulations, *FAS-FAX*.

14. *Wired* 1995 Advertising Kit.

15. *Get Wired* Advertising Kit, 7.

16. For example, the March 1994 issue of *Wired* contains 51 pages of advertising and advertorials out of a possible 146 pages, while the March 1995 issue has 68 out of a possible 174.

17. *Get Wired* Advertising Kit, 15.

18. Susan Countryman and Suzanne Galante, "Condé Nast Nabs Wired," News.com. <http://www.news.com/New/Item/0,4,21923,00.html>. May 8, 1998.

19. See The Global Business Network Web site: <http://www.gbn.org/home.html>.

20. Borsook, "The Memoirs of a Token: An Aging Berkeley Feminist Looks at *Wired*," 30.

21. Ibid., 36.

22. Simone de Beauvoir's *The Second Sex* remains one of the best explorations of how these myths have functioned to define women as Other.

23. There is little doubt that masculinity is *Wired*'s main focus. However, because this book is a feminist project that seeks to place women at the centre of the analysis, I have chosen to prioritize women in the order of the next two chapters. This choice prevents my work from replicating *Wired*'s subordination of women through a complementary presentation and emphasizes the fact that how digital discourse constructs difference is just as important to understanding *Wired*'s masculine appeal as it is for understanding how it constructs its dominant masculine ideal.

Chapter Four

1. Constance Hale, ed., *Wired Style: Principles of English Usage in the Digital Age* (San Francisco: HardWired Books, 1996), 2.

2. Ray Bradbury, "Triangulation," *Wired*, August 1997, 92.

3. Nicholas Negroponte, "Bits and Atoms," *Wired*, January 1995, 176.

4. John Heilemann, "Do You Know the Way to Ban Jose?," *Wired*, August 1996, 45–46, 48, 176–181.

5. Geremie R. Barme and Sang Ye, "The Great Firewall of China," *Wired*, June 1997, 138–151, 182.

6. Po Bronson, "The Relentless Pursuit of Connection," *Wired*, November 1995, 11–12.
7. Hale, ed., *Wired Style*, 95.
8. From Pamela McCorduck, "Sex, Lies and Avatars," *Wired*, April 1996, 13–16.
9. Clive Davidson, "Christine Downton's Brain," *Wired*, December 1996, 170.
10. Hari Kunzru, "You Are Borg," *Wired*, February 1997, 156.
11. Lycos advertisement, *Wired*, July 1998, 145.
12. Phil Patton, "Dream Ware," *Wired*, April 1994, 96.
13. Claudia Springer, *Electronic Eros: Bodies and Desire in the Postindustrial Age* (Austin: University of Texas Press, 1996), 9.
14. See Marshall McLuhan, *The Mechanical Bride: The Folklore of Industrial Man* (Boston: Beacon Press, 1967), 94.
15. Cover, *Wired*, November 1997.
16. "Special Report: Hollywood 2.0," *Wired*, November 1997, 201.
17. Richard Kadrey, "(Im)material Girl," *Wired*, March 1997, 51.
18. Sega Saturn advertisement, *Wired*, November 1995, 106–107.
19. Mary White Stewart, "A Visual Essay on Women in *Vogue* 1940-1990," in V. Bentx and P. Mayes, eds., *Women's Power and Roles as Portrayed in Visual Images of Women in the Arts and Mass Media* (Lewiston: Edwin Mellen Press, 1993), 19–36.
20. Mary Elizabeth Williams, "Hack and Ye Shall Learn," *Wired*, December 1996, 69.
21. See *Wired*, April 1994 and May 1995.
22. OS/2 Warp, IBM advertisement, *Wired*, March 1995, 4–5.
23. Kevin Kelly, "Manners Matter," *Wired*, November 1997, 224–225.
24. Gilbert Paper advertisment, *Wired*, September 1994, 5–6.
25. Kingston Technology advertisement, *Wired*, January 1996, 79.
26. See, for example, "The Daedalus Encounter," Virgin Interactive Entertainment advertisement, *Wired*, June 1995, 124–125; and "Noctropolis," Electronic Arts advertisement, *Wired*, December 1994, 75.
27. Squaresoft advertisement, *Wired*, February 1997, 177.
28. Play advertisement, *Wired*, June 1997, 29.
29. See *Wired*, December 1994, 20; May 1995, 53; August 1995, 35; September 1995, 46; and April 1996, 39, respectively.
30. "Quake Girls," *Wired*, April 1996, 39.
31. *Wired*, November 1995.
32. *Wired*, February 1995, 119.
33. Cover and Michelle Slatalla and Joshua Quittner, "Gang War in Cyberspace," *Wired*, December 1994.
34. Steve G. Steinberg, "Gigabit Networker," *Wired*, May 1998, 87.
35. Matt Haber, "RapDotCom," *Wired*, November 1995, 168.

36. Mark Frauenfelder, "Afroamerica Online," *Wired*, September 1995, 43.
37. Lucent Technologies advertisement, *Wired*, January 1998, 58–59.
38. It is also reminiscent of the work of communications guru Marshall McLuhan, a Canadian theorist who used text and image to portray the future of telecommunications in his and Quentin Fiore's seminal work, *The Medium Is the Message* (New York: Random House, 1967).
39. Packard Bell advertisement, *Wired*, December 1996, 19–20.
40. Acer advertisement, *Wired*, May 1996, 75.
41. Ziauddin Sardar, "alt.civilizations.faq: Cyberspace as the Darker Side of the West," in Z. Sardar and J. R. Ravetz, eds., *Cyberfutures: Culture and Politics of the Information Superhighway* (New York: New York University Press, 1996), 30.
42. The term "fembot" was coined by Mary Daly to describe subservient women. See Mary Daly, *Gyn/Ecology: The Metaethics of Radical Feminism* (Boston: Beacon Press, 1990), 17.

Chapter Five

1. Ellen McCracken, *Decoding Women's Magazines: From Mademoiselle to Ms.* (New York: St. Martin's Press, 1993), 13.
2. See *Wired* covers, November 1995, March 1996, August 1996, February 1995 and October 1996, respectively.
3. Cover, *Wired*, October 1996.
4. Cover, *Wired*, October 1995.
5. Cover, *Wired*, September 1995.
6. Cover, *Wired*, June 1997.
7. Constance Hale, ed., *Wired Style: Principles of English Usage in the Digital Age* (San Francisco: HardWired Books, 1996), 2.
8. Pamela McCorduck, "Sex, Lies and Avatars," *Wired*, April 1996, 106–111, 158–165; Hari Kunzru, "You Are Borg," *Wired*, February 1997, 155.
9. Hale, ed., *Wired Style*, 2, 7.
10. Roderick Simpson, "Privacy by Geometry: Elliptic Curves and Low Cost-Per-Bit Crypto Strength," *Wired*, December 1997, 112.
11. NEC advertisement, *Wired*, February 1997, 28–29.
12. IBM (International Business Machines); MS (Microsoft); HP (Hewlett Packard); NSF (National Science Foundation); FCC (Federal Communications Commission); ASCII (American Standard Code for Information Interchange); BASIC (Beginner's All-Purpose Symbolic Instruction Code); CPU (Central Processing Unit); DOS (Disk Operating System); LAN (Local Area Network); HTML (Hypertext Markup Language); PPP (Point to Point Protocol); SLIP (Serial Line Internet Protocol).
13. Gareth Branwyn, ed., "Jargon Watch," *Wired*, March 1996, 56.
14. Gareth Branwyn, ed., "Jargon Watch," *Wired*, April 1997, 68.

15. Roger Fowler, "Power," in Teun van Dijk, *Handbook of Discourse Analysis,* Volume 4 (London: Academic Press, 1985), 61–82.
16. John A. Barry, *Technobabble* (Cambridge: MIT Press, 1991), 5.
17. Ibid.
18. Bart Kosko, as quoted by Sheldon Teitebaum, "Fuzzy Logic," *Wired,* February 1995, 135.
19. Fowler, "Power," 69.
20. Greg Blonder, "Faded Genes," *Wired,* March 1995, 107.
21. Jon Katz, "Return of the Luddites," *Wired,* June 1995, 162.
22. Steven Levy, "Insanely Great: Ode to an Artifact," *Wired,* February 1994, 58.
23. Kevin Kelly, "A Globe Clothing Itself in a Brain," *Wired,* June 1995, 108-115.
24. David Kline, "Align and Conquer," *Wired,* February 1995, 110–118.
25. Penn Jillette, as quoted by Joshua Quittner, "Penn," *Wired,* September 1994, 97.
26. Jeff Goodell, "The Supreme Court," *Wired,* March 1995, 116–122.
27. Bob Johnstone, "True Boo-Roo," *Wired,* March 1995, 137.
28. Michelle Slatalla and Joshua Quittner, "Gang War in Cyberspace," *Wired,* December 1994, 148.
29. James der Derian, *Wired,* September 1994, 116–124, 158 and April 1995, 137–138.
30. Sony PC advertisement, *Wired,* December 1995, 5–6.
31. Lotus Development advertisement, *Wired,* May 1997, 49.
32. Tim Barkow, "Bottom Feeders," *Wired,* October 1996, 110.
33. Levy, "Insanely Great: Ode to an Artifact," 58.
34. Phil Patton, "Dream Ware," *Wired,* April 1994, 96–98.
35. Blue Sky Entertainment advertisement, *Wired,* January 1995, 25.
36. Hillary Rettig, "The King of Quant," *Wired,* March 1995, 86; Michael Schrage, "Revolutionary Evolutionist," *Wired,* July 1995, 120; and Evan I. Schwartz, "The Father of the Web," *Wired,* March 1997, 140.
37. Gary Wolf, "The Wisdom of Saint Marshall, Holy Fool," *Wired,* January 1996, 123, 128.
38. Katie Hafner, "The Great Creators," *Wired,* December 1994, 152.
39. "Infobahn Warrior" cover headline refers to feature interview by David Kline, "Big Bad John," *Wired,* July 1994, 86–90, 130–131; "The Cable Slayer" cover headline refers to feature interview by David Kline, "Align and Conquer," *Wired,* February 1995, 110–117, 164; and John Schwartz, "Angriest Guy in All of Cyberland," *Wired,* April 1995, 47.
40. David Kline, "Align and Conquer," *Wired,* February 1995, 164; Joshua Quittner, "Penn," *Wired,* September 1994, 101; Michael Schrage, "Revolutionary Evolutionist," *Wired,* July 1995, 123.
41. Dorion Sagan, "Sex, Lies and Cyberspace," *Wired,* January 1995, 78–84.

42. Joshua Quittner, "Automata Non Grata," *Wired*, April 1995, 118–122; Robert Rossney, "Doom from Above," *Wired*, January 1996, 176.
43. Paramount Interactive advertisement, *Wired*, April 1994, 23.
44. Netrunner advertisement, *Wired*, November 1996, 131.
45. 3D Realms Entertainment advertisement, *Wired*, December 1996, 177.
46. "Noctropolis," Electronic Arts advertisement, *Wired*, December 1994, 75.
47. Brock N. Meeks, "Fuelling the 'Net Porn' Hysteria," *Wired*, September 1995, 80.
48. Mike Godwin, "Net Backlash = Fear of Freedom," *Wired*, August 1995, 70.
49. Brock N. Meeks, "The Obscenity of Decency," *Wired*, June 1995, 86.
50. Brock N. Meeks, "Fuelling the 'Net Porn' Hysteria," *Wired*, September 1995, 80.
51. Zackary Margulis, "Canada's Thought Police," *Wired*, March 1995, 92.
52. Bob Johnstone, "Godzone," *Wired*, November 1995, 164–167, 230.
53. David Kline and Daniel Burnstein, "Is Government Obsolete?," *Wired*, January 1996, 86–108.
54. Kevin Kelly, "Interview with the Luddite," *Wired*, June 1995, 166–168, 211.
55. Harvey Blume, "Digital Refusnik," *Wired*, May 1995, 178–179.
56. Jon Katz, "Return of the Luddites," *Wired*, June 1995, 164.
57. Ibid., 165.
58. Stuart Brand, "Paglia," *Wired*, January 1993. The interview is only available online at <http://www.wired.com/wired/1.1/features/paglia.html>; Rosie Cross, "Modem Grrl," *Wired*, February 1995, 118–119.
59. See Zackary Margulis's portrayal of Catharine MacKinnon, for example, in "Canada's Thought Police," *Wired*, March 1995, 93.
60. Margulis, "Canada's Thought Police," 92.
61. MacKinnon's approach to pornography does not, of course, represent the only feminist position, despite *Wired*'s simplistic characterization. In fact, the issue of censorship and pornography is one of the most widely debated issues in feminist theory.
62. Digital Pictures advertisement, *Wired*, January 1996, 5, 6.
63. Maniac Sports CD-Rom advertisement, *Wired*, August 1994, 65.

Chapter Six

1. Steven Levy, "Insanely Great," *Wired*, February 1994, 58.
2. Robyn Rand, as quoted by Jon Carroll in "Guerilla's in the Myst," *Wired*, August 1994, 73.
3. Levy, "Insanely Great," 58.
4. Phil Patton, "Dreamware," *Wired*, April 1995, 95.
5. Charles Platt, "What's It Mean to Be Human Anyway?," *Wired*, April 1995, 132.
6. Greg Blonder, "Faded Genes," *Wired*, March 1995, 107.

7. Kristin Spence, "The Biggest Little Lab in the World," *Wired*, September 1995, 190.
8. Oliver Morton, "Overcoming Yuk," *Wired*, January 1998, 44.
9. Jon Katz, "The Return of the Luddites," *Wired*, June 1995, 162.
10. Michael Schrage, "Revolutionary Evolutionist," *Wired*, July 1995, 120.
11. Canada, *Final Report of the Information Highway Advisory Council, Connection, Community, Content: The Challenge of the Information Highway* (Ottawa: Industry Canada, 1995).
12. Erik Davis, "Technopagans," *Wired*, July 1995, 126.
13. Ibid.
14. Ed Regis, "Extropians," *Wired*, October 1994, 104.
15. Digital discourse has also explored possibilities of moving cybernetic devices into the real world. The dream is to construct a world in which objects can respond and interact in order to meet human needs. Human and machine would ostensibly live in harmony. See Mark Slouka, *War of the Worlds: Cyberspace and the High Tech Assault on Reality* (New York: Basic, 1995), 66–68.
16. IBM OS/2 Warp advertisement, *Wired*, January 1995, 7–8.
17. TDK advertisement, *Wired*, February 1995, 33.
18. Hooked.inc. advertisement, *Wired*, February 1995, 37.
19. Worldwide Internet Publishing Corporation advertisement, *Wired*, March 1998, 197.
20. John A. Barry, *Technobabble* (Cambridge: MIT Press, 1991), 65–66.
21. David Rothenberg, "Deep Technology," *Wired*, October 1995, 121.
22. Ed Regis, "The Environment is Going to Hell and Human Life is Doomed to Only Get Worse, Right? Wrong. Conventional Wisdom, Please Meet Julian Simon, the Doomslayer," *Wired*, February, 1997, 137–139.
23. John E. Young, *Global Network: Computers in a Sustainable Society* (Washington: Worldwatch Institute, 1993), 37–38.
24. Daryl Fields and Jack Ruitenbeek, *Sustainability and Prosperity: The Role of Infrastructure* (Ottawa: National Round Table on Environment and Economy: Institute for Research on Public Policy, 1992), and Third World Network, "Modern Science in Crisis: A Third World Response," in Sandra Harding, ed., *The Racial Economy of Science: Toward a Democratic Future* (Bloomington: Indiana University Press, 1993), 512.
25. Young, *Global Network*, 38–47.
26. It is important to recognize here that "nature" in western culture is itself seen through the lens of culture and technology. I do not mean to suggest that "nature" and our conception of it is purely a-cultural. There is little doubt that western perceptions of nature, even in ecological movements, are highly influenced by how nature is constructed in television nature programs and so on. This fact is sometimes obscured by environmentalists who invoke images of a "pristine" natural

world, unspoiled by human hands and unseen by human eyes — a world that for all intents and purposes never has existed (and will never exist). The goal here is not to replace the myth of cyberspace-as-Garden-of-Eden with a myth of nature-as-Garden-of-Eden, but to recognize that the objectification and commodification of nature has serious implications for the long-term sustainability of the planet and for our perceptions of our world and of one another.

27. Art Kleiner, "The Battle for the Soul of Corporate America," *Wired*, August 1995, 125.
28. Cyberfeminist Sadie Plant echoes this theme, arguing that digital technologies are creating a world beyond anything we can possibly imagine, "as if history was merely the prehistory of cyberfeminism." "Beyond the Screens: Film Cyberpunk and Cyberfeminism," *Variant* 14 (1993), 17.
29. For a discussion of how western corporations are engaged in a process of patenting living organisms, see Vandana Shiva, *Biopiracy: The Plunder of Nature and Knowledge* (Toronto: between the lines, 1997).
30. Bill Gates, as quoted in *Business Week*, June 27, 1994, in Ed Regis, "Hacking the Mother Code," *Wired*, October 1995, 137.
31. Michael Schrage, "Revolutionary Evolutionist," *Wired*, July 1995, 121.
32. These are clearly the rights that corporations seek to obtain through the Multilateral Agreement on Investment (MAI), currently being negotiated through the Organization for Economic Cooperation and Development (OECD) and through Trade Related Intellectual Property Rights (TRIPs) as part of the General Agreement on Tariffs and Trade (GATT).
33. Ziauddin Sardar and Jerome R. Ravetz, "Introduction: Reaping the Technological Whirlwind," *Cyberfutures: Culture and Politics On the Information Highway* (New York: New York University Press, 1996), 2.
34. Jeremy Rifkin, *The Biotech Century: Harnessing the Gene and Remaking the World* (New York: Jeremy P. Tarcher/Putnam, 1998), 51–52.
35. Despite all this public assistance and *Wired*'s rhetoric about cybercommunity, Paulina Borsook notes that information technology companies and professionals are unwilling to return the favour. In fact, high-tech industry professionals are among the least likely to support charities in the United States, even though they are among the most financially able to do so. See Paulina Borsook, "Cyberselfish," *Mother Jones*, July/August 1996, 59.
36. Theodore Roszak describes this phenomenon as "the cult of information," in *The Cult of Information* (Berkeley, CA: University of California Press, 1994).
37. Ziauddin Sardar, "alt.civilizations.faq: Cyberspace as the Darker Side of the West," in Sardar and Ravetz, eds., *Cyberfutures: Culture and Politics On the Information Highway*, 29–30.
38. Ibid., 26.
39. Nyla Ahmad, "CyberSurfer," *Owl Magazine*, April 1996, 20–23.
40. Corina M. Koch, "Is Free-Time Fair? Girls, Boys and Classroom Computer Use,"

Professionally Speaking (March 1998), 40–42.

41. Roszak, *The Cult of Information*, 88.
42. C. A. Bowers, *The Cultural Dimensions of Educational Computing: Understanding the Non-Neutrality of Technology* (New York: Teachers College Press, 1988).
43. See, for example, Kevin Kelly, "Brian Eno: Gossip is Philosophy," *Wired*, May 1995, 148–149.
44. Roger Bertelson, factory manager for Motorola, as quoted by William Greider in *One World, Ready or Not: The Manic Logic of Global Capitalism* (New York: Simon and Schuster, 1997), 83.
45. *Wired*, January 1995, 163.
46. Erik Davis, "Digital Dharma," *Wired*, August 1994, 54–59.
47. A. Lin Neumann, "The Resistance Network," *Wired*, January 1996, 108.
48. Neal Stephenson, "In the Kingdom of Mao Bell or Destroy the Users on the Waiting List," *Wired*, February 1994, 86.
49. Bob Johnston, "Wiring Japan," *Wired*, February 1994, 38.
50. John Andrew, "Culture Wars," *Wired*, May 1995, 138.
51. Microsoft's current industry dominance may be only a fraction of what is to come. The company's interest in developing the potential of interactive television in order to expand its market is well known. Investments in biotechnology are also likely to figure more heavily in the company's future. If Microsoft's pattern of achieving industry dominance in those sectors it enters continues, diverse information systems world wide may well bear the Microsoft trademark.
52. A good example is provided by Nicholas Negroponte's "Why Europe Is So Unwired," *Wired*, September 1994, 160. Negroponte argues that the United States is destined to lead the digital revolution because of "our venture capital system" which rewards entrepreneurs willing to take risks without being "thwarted by the bureaucracies of a homogenous, old society" concerned about cultural protection.
53. Marike Finlay, *Powermatics* (London: Routledge and Kegan Paul, 1987), 30. Italics in original.
54. This is discussed in detail in David Noble, *Progress Without People: New Technology, Unemployment and the Message of Resistance* (Toronto: between the lines, 1995).
55. *The Road Ahead* is the title of the book written by Microsoft Corporation CEO, Bill Gates (New York: Penguin, 1996).

Chapter Seven

1. Irshad Manji, *Risking Utopia: On the Edge of a New Democracy* (Vancouver: Douglas and McIntyre, 1997), 150.
2. CBC Radio, *Ideas*, "Luddites and Friends," Part II, November 12, 1997.
3. Jim Carroll, *Surviving the Information Age* (Scarborough, ON: Prentice-Hall Canada, 1997), 5.

4. Donald Tomaskovic-Devey, *Gender and Racial Inequality at Work: The Sources and Consequences of Job Segregation* (Ithaca, NY: ILR Press, 1993), 149.
5. Ibid., 173.
6. Ibid., 15.
7. Swasti Mitter, "Beyond the Politics of Difference: An Introduction," in Swasti Mitter and Sheila Rowbotham, eds., *Women Encounter Technology: Changing Patterns in Employment in the Third World* (London: Routledge, 1995), 17.
8. Lynne Segal, "New Introduction to the 1997 Edition," *Slow Motion: Changing Masculinities Changing Men*, rev. ed. (London: Virago Press, 1997), xxxii.
9. Heather Menzies, *Whose Brave New World? The Information Highway and the New Economy* (Toronto: between the lines, 1996), 159–162. The Tobin tax was first proposed in 1972 by economist James Tobin. The 1981 Nobel Prize winner's call for a small levy on international transactions to be applied at major foreign-exchange centres has sparked more interest in the 1990s than during the booming economic times of the 1970s. Today, supporters believe that such a tax would not only slow the intense pace of global finance, but would also yield billions in revenue that could be reinvested in social welfare programs and used to achieve a greater degree of economic equality.
10. Motorola Multimedia Group advertisement, *Wired*, November 1995, 65.
11. Vida Panitch, "From Bread Riots to Cyber-Revolution," *Women'space* 3, no. 2 (Fall 97/Winter 98), 11.
12. See <http://www.riotgrrl.com/>.
13. See <http:// www.cybergrrl.com/>.
14. Aliza Sherman, *Cybergrrl! A Woman's Guide to the World Wide Web* (Ballatine Books, 1998). Book information, complete with a link to order a copy, can be found at <www.cybergrrl.com/planet/cgbook/book.html>.
15. Illustration by Juliet Breese, *Women'space* 3, no. 2 (Fall 97/Winter 98), cover.
16. Ruth Schwartz Cowan, *More Work for Mother: The Ironies of Household Technology from the Open Hearth to the Microwave* (London: Free Association Books, 1989).
17. In the words of Jennifer Light, "engendering new uses degenders the computer." See "The Digital Landscape: New Space for Women," *Gender, Place and Culture* 3, no. 2 (1995), 134.
18. The online group Women Halting Online Abuse (WHOA) provides some excellent step-by-step information about what women can do to confront online sexual harassment. See <http://www.whoa.femail.com/harass.html>.
19. Susan Herring, "Posting in a Different Voice: Gender and Ethics in Computer-Mediated Communication," in Charles Ess, ed., *Philosphical Perspectives on Computer-Mediated Communication* (Albany, NY: State University of New York Press, 1997), 115–147.
20. Eric von Hippel, as cited in C. Fischer, "The Telephone Industry Discovers Sociability," *Technology and Culture* 29 (1988), 32–61.
21. This raises many complex questions about whether or not the "intrusion" of digital

equipment into our homes will affect the public/private divide that has served to disadvantage women in the past. While it is clear that this issue requires more investigation, it would seem unlikely that simply moving work technology into the home will result in significant changes in the power relations of gender.

22. Covers, *The Globe and Mail Report on Business Magazine*, April 1998 and May 1998, respectively.

Glossary

Chatroom: An online forum that allows Internet users in different locations to "talk" to each other electronically in real time.

Class: A social division based on economic status. Class differences in capitalist societies are a perpetual source of injustice and conflict. Class analysis allows us to better understand the power relations of our society by revealing these socioeconomic disparities.

Cracker: A term used in digital culture to refer to an outsider who breaks into (or cracks) computer systems without authorization. A cracker is different from a hacker, who may or may not use her/his computer expertise to gain access to secured computer systems. This books seeks to crack the gender code used in digital culture.

Cyberfeminism: A women-centred perspective that advocates women's use of new information and communications technologies for empowerment. Some cyberfeminists see these technologies as inherently liberatory and argue that their development will lead to an end to male superiority because women are uniquely suited to life in the digital age.

Cyberspace: Coined by William Gibson in his 1984 science fiction novel *Neuromancer*, the term refers to the "space" between computers where online information is found and retrieved. "Cyberspace" has become a dominant metaphor of digital culture.

Cyborg: A cybernetic organism that is part-machine, part-human. The term has been popularized in academic feminism by Donna Haraway. She argues that because of the degree to which technology has infused our lives and selves in the late twentieth century, "we are all cyborgs."

Digital, digital technology: Digital technology is used to store information in digital form, as a string of separate bits that represent on/off states. Digital code is often described and represented as a string of ones and zeroes. Digital technology is significant because it has allowed diverse forms of information to be easily stored and transported electronically.

Digital culture: The customs, lifestyle and shared values that are emerging around digital technologies. Digital culture is reflected in its most intense form in discourses like that found in *Wired* magazine and in the corporate and scientific world of digital technology. However, as these technologies and their accompanying discourse and ideology spread, mainstream western culture is itself becoming increasingly digital in nature.

Digital discourse: The specific language, text, talk, symbols and shared meanings that surround digital technologies and the new information economy. There are many forms of digital discourse in our society, from the most intense forms (as found in magazines like *Wired*) to that used by journalists and advertisers in mainstream culture.

Digital generation: Defined by *Wired* magazine as the people who read *Wired* and who are "the change leaders" of our time. According to a 1997 *Wired* Subscriber Survey, the average member of the digital generation is a 39-year-old male college graduate with an annual income of US$83,000. Members of the digital generation are powerful individuals because they are employed in high status positions in business, science and technology. As a result, they make many important decisions about how the new information economy will be structured.

Digital ideology: The beliefs, values and political dispositions that are emerging around new information and communications technologies. Digital ideology is characterized by an acute faith in speed, technology and technological progress, political libertarianism and corporate imperialism. It also incorporates sexism and racism, scientific reductionism and environmental exploitation.

Discourse: A language structure that includes particular words and jargon, symbols, metaphors, meanings and associations. Our society is filled with many different discourses that have varying degrees of specificity, ideological content and power. For example, doctors and nurses use a particular medical discourse in order to describe and understand the human body and their profession. Inherent in this discourse are a number of assumptions about health, disease, normalcy and western science.

Discursive: Of or having to do with discourse as opposed to material conditions. Relations of power can be described as discursive.

Domestic (or private) sphere: The "home," associated with the feminine sphere in western culture and rigidly separated from the public sphere of paid work and political institutions. Many feminists view the exploitation of women in the private sphere as fundamental to the dynamics of both capitalism and patriarchy.

Electronic Frontier Foundation (EFF): An online civil liberties organization founded in 1990 by Americans Mitch Kapor and John Perry Barlow. The EFF is primarily concerned with ensuring that the internet is free from government regulation and censorship. Former chairman Esther Dyson argues that the EFF is shifting its focus to deal less with ensuring that government stay out of cyberspace and more with devising appropriate forms of self-governance.

Enlightenment: A period of intellectual change in eighteenth-century European societies. During the Enlightenment, human reason attained superiority over superstition and religion, and a belief in progress became dominant.

Femininity: A collection of gender characteristics associated with women, which have been subject to historical evolution. In western culture, "feminine" is associated with inferiority, weakness, irrationality, beauty, nature, reproduction and immanence. Feminism has sought to expand and destabilize hegemonic definitions of femininity in order to liberate women from patriarchal constructions of femininity that limit women's freedom to define themselves.

Gender: The group of socially constructed characteristics ascribed to each sex. While an individual's "sex" refers to more narrowly defined biological traits (reproductive organs, x and y chromosomes), an individual's "gender" is more broadly defined by social interactions and assumptions about femininity and masculinity. Although the distinction between "sex" and "gender" is itself the artificial product of western scientific culture, separating the two has allowed feminists to dispute patriarchal claims that "biology is destiny." Simone de Beauvoir emphasized the significance of gender over sex in *The Second Sex* when she argued that "one is not born, but rather becomes a woman."

Gender code: The underlying beliefs and assumptions about masculinity and femininity that are promoted by digital culture. This code is digital not only because it is associated with digital culture but also because it is often rigidly binary in nature. Gender codes are a form of social control that influence how men and women perceive themselves and their world.

Hegemonic: Dominant; often used to refer to views that dominate our society and are generally accepted. The term is associated with the work of Italian political theorist Antonio Gramsci. Gramsci used the concept to describe how the ruling capitalist class maintains their power over the populace by manipulating the process of socialization to reflect their own values and interests.

Hypermacho man: The quintessential masculine ideal found in *Wired* magazine and increasingly evident throughout popular culture. The hypermacho man exhibits an intensified and accelerated form of machismo intimately associated with his close relationship to high technology.

Hypermodernity: The intensified, accelerated form of modernity found in digital discourse and, increasingly, in much of contemporary mainstream culture.

Immanence: Inherent, of this world, associated with survival and the sustenance of human life. In western patriarchal culture, immanence is seen as inferior to transcendence and is defined as a feminine preoccupation.

Information economy: Distinguished from the older factory-oriented economy by the centrality of information as the primary product. The information economy is global rather than national in scale and is facilitated by the development of new information and communications technologies that provide the high speed international networks it requires.

Internet: A worldwide network of computers linked together, most commonly, by telephone lines. The internet evolved from ARPANET, a communications project of the U.S. Defense Department.

Libertarianism: An extreme form of anti-government, pro-free market ideology espoused by many of the digital generation. Digital libertarians seek complete freedom from the restraint of government regulation and frequently focus on the dangers of censorship as a way of supporting a completely privatized Internet. Central to libertarianism is the belief in the sanctity of individual rights, particularly the right of the individual to own and accumulate property without interference.

List-serv: An online service that allows users to subscribe to specific mailing lists. List-servs allow subscribers to receive electronic mail from other users interested in the same topic or issue.

Luddites: Members of a political movement of the Industrial Revolution in the early-nineteenth century. Named after their leader, Ned Ludd, the Luddites were largely hand weavers who opposed the mechanization of their craft, sometimes by violently "smashing" the new machines. Despite the fact that the original Luddites were not opposed to technology as such but to the economic implications of specific uses of technology, the term "Luddite" has become a derogatory term used to describe anyone who is deemed to be "antitechnology."

Masculinity: A collection of gender characteristics associated with men. The dominant form of masculinity has, like femininity, been subject to historical evolution. Masculinity is generally associated with superiority, strength, reason, science and transcendence. Like constructions of femininity, hegemonic masculinity restricts the ways men can define themselves and imposes rigid limits on behaviour and self-actualization.

Material: Concrete economic and structural conditions that affect our lives. The "material conditions of women's lives" includes the many institutional and political barriers (which can be sexist, racist, classist) that women face on a daily basis.

Modem: A device that allows computers to receive and transmit information across ordinary telephone lines. A "modulator-demodulator" and a computer are required to access online resources and send and receive e-mail. St. Jude's notion that "girls need modems" expresses the cyberfeminist belief that new information and communications technologies are important tools for women's liberation.

Modern, modernity: Used to refer both to a social reality (modern society) and to an intellectual movement (modern thought). Originating during the Enlightenment, modernity is associated with the attempt to use scientific reason to order the social and natural world. Modernity is closely linked both to the rise of western industrial capitalist society and to the elevation of science and technology. Modern thinkers seek to overcome the chaos and rapid change of modern society by imposing rational, scientific order on the world.

MUDs: Multi-user dungeons or domains; virtual spaces where participants create artificial online communities and identities through the text they type. MUDs had their origins in Dungeons and Dragons, a role-playing board game. MOOs are similar, except they are "object oriented," or dedicated to a specific goal or task.

Multilateral Agreement on Investment (MAI): Proposed treaty that has been under negotiation amongst member countries of the Organization for Economic Co-operation and Development (OECD). While negotiations are currently suspended, they may be resumed in the future. The treaty is likely to include provisions to enhance the mobility of capital investment and limit government regulation of foreign investment. Seen as an extension of existing free trade agreements, the MAI is viewed by critics as imposing further restrictions on the sovereignty of nation–states, while enhancing corporate power.

Newsgroup: A public discussion area on the Internet where participants can post and read messages of interest to other members. Newsgroups are different from list-servs because users have to "visit" the area to read messages rather than automatically receive messages at their e-mail address.

Online: Connected, "wired," having the ability to access Internet material and communications using a computer and modem. Getting online is a prerequisite to any form of participation in the digital age.

Other: As opposed to the "self," the Other is a term used to represent "inferior" non-white, non-male or non-western human beings. The self/other dualism is a crucial hierarchical distinction in western theory and culture.

Popular culture: The mainstream and widely accessible knowledge, customs, folklore and arts of a given society. Popular culture includes television, movies, music, art, print media, advertising and so on created for a mass audience. American popular culture has become ubiquitous in the globalized post-television world of the 1990s.

Postmodern: Used to refer both to a social reality (postmodern society) and to an intellectual movement (postmodern thought). Postmodernity arose during the late 1960s and early 1970s as a response to the crisis of modernity. Postmodern society is characterized by great speed, fragmentation and the elevation of representation over reality. Postmodern theory attempts to overcome the universal totalizing theory of modernity by deconstructing its grand narratives and unified subject and embracing diversity. One of the central insights of postmodern thought is the significance of discourse to the construction of social realities and identities.

Power relation: A term used to describe the many incidents of domination and subordination found in society. Multiple power relations flow throughout society. Hegemonic power relations create many easily recognizable hierarchies, including men over women, white over black, heterosexual over homosexual and so on.

Public sphere: The world of government and business, associated with men and distinguished from the unpaid domestic or private sphere of women and children.

Redeploy: To redistribute, re-circulate or reproduce cultural images or messages that are associated with the past and that are regressive in nature. Digital discourse redeploys sexist images of women that are reminiscent of 1950s cultural norms.

Reductionism: The oversimplification and breaking down of complex phenomena into discrete components. Scientific reductionism may have the unfortunate result of viewing life as the sum of its parts and ignoring the intrinsic worth, unity and interconnectedness of organisms.

Second-wave feminism: Linked to the Women's Liberation Movement of the 1960s and 1970s and distinguished from the first wave of feminism that was closely associated with the suffragist movement of the late 1800s. Feminists of the second wave are perhaps best known for the notion that "the personal is political," a slogan that recognizes the multiple levels of women's oppression and validates women's personal experiences as the foundation for political struggle.

South: Countries of the southern hemisphere, many of which are also referred to as "developing," "Third World" or "disadvantaged" nations. These nations have been subject to a long history of exploitation at the hands of industrialized countries of the North, which are defined as the industrial nations of North America and Europe.

Technobabble: The language and jargon used to describe specific technologies. Technobabble is often difficult for outsiders and nontechnical people to understand and is frequently characterized by a high number of acronyms and initialisms.

Technological determinism: The idea that technology determines society. Technological determinism mystifies the fact that technologies are designed within specific social and economic structures by individuals and suggests that social development is driven by inevitable technological progress. The term also imposes a distinction between technology and society that is difficult to sustain; technology is, after all, intimately related to society, and vice versa.

Technology: Generally defined as the practical or industrial arts; the application of science to meet human needs. Alternative definitions of technology broaden this definition to include all the practices and knowledge that humanity uses, whether in the public or the private sphere. Such alternative definitions also highlight the fact that technology is always embedded within larger social power relations.

Technomanic: Literally, a form of technology-related madness. The term is used to describe views (and people) that are wildly enthusiastic about technology and its ability to transform our world for the better.

Technophilic: Views (and people) that express a "love" and devotion for technology. Technophiles often express functional, optimistic views of the role of technology in society.

Technophobic: Views (and people) that express (sometimes irrational) fears or aversions to technology; people who are afraid of using new technologies. Technophobes often express negative views about the role of technology in society.

Technotopia: The futuristic fantasy world that will emerge once technology eliminates all social, economic and political problems. While such a world is highly improbable, the myth of an imminent technotopia is a useful way of legitimizing our technologically driven society.

Technovangelism: A form of pseudo-religious preaching designed to convert others to believe in the wonders of technology. Technovangelism often uses explicit religious metaphors in order to elevate the status of technology as a means of achieving divine transcendence. *Wired* magazine includes many examples of technovangelism as it tries to sell a sort of digital Nirvana of the future.

Transcendence: The goal of rising above or going beyond the simple sustenance of human life in order to attain deeper meaning, truth or perfection. Transcendence is sought in a number of ways: through the pursuit of philosophical wisdom, religion, art, literature and, more recently, through scientific reason and technological progress. In western patriarchal culture, transcendence is defined as superior to immanence and is seen as a masculine pursuit.

Virgin/whore dichotomy: A long-standing dualism in western, patriarchal culture in which women are viewed either as pure, innocent "virgins" or evil, highly sexualized "whores." This dualism is perpetuated in digital discourse through the construction of cyber-Barbies and cyber–femme fatales.

Virtual reality (VR): Used to describe computer-generated environments designed to simulate real environments in a very lifelike way. The success of virtual reality technologies is measured by its closeness to "real life."

Web sites: Specific online addresses or locations on the World Wide Web that display specific information; a collection of Web pages that users "visit" to access information or services. Web sites feature a great variety of information, some of which is useful for feminist research and networking.

Wired world: The connected, online social reality created by the new digital technologies and culture. The wired world includes both the virtual world of cyberspace and the real life digital culture and economy that surrounds it.

World Wide Web: The most accessible and graphic incarnation of the Internet. Web sites allow users to "click" on hyperlinks to instantly move from one site to another.

Zines: (Pronounced zeenes.) Low-budget magazines published independently for a small niche market. E-zines are published electronically for an online audience.

Selected Bibliography

Adas, Michael. *Machines as the Measure of Men: Science, Technology and Ideologies of Western Dominance.* Ithaca: Cornell University Press, 1989.

Balsamo, Anne. *Technologies of the Gendered Body: Reading Cyborg Women.* Durham, NC: Duke University Press, 1996.

Bender, Gretchen, and Timothy Druckrey, eds. *Culture on the Brink: Ideologies of Technology.* Seattle: Bay Press, 1994.

Brook, James, and Iain A. Boal, eds. *Resisting the Virtual Life: The Culture and Politics of Information.* San Francisco: City Lights, 1995.

Cameron, Deborah, ed. *The Feminist Critique of Language: A Reader.* London: Routledge, 1990.

Castells, Manuel. *The Rise of the Network Society.* Malden, MA: Blackwell, 1996.

Cherny, Lynn, and Elizabeth Reba Weise, eds. *wired_women: Gender and New Realities in Cyberspace.* Seattle: Seal Press, 1996.

Chodos, Robert, Rae Murphy, and Eric Hamovitch. *Lost in Cyberspace? Canada and the Information Revolution.* Toronto: James Lorimer and Co., 1997.

Cushing, Pamela J. "Gendered Conversational Rituals on the Internet: An Effective Voice Is Based on More than Simply What One Is Saying." *Anthropologica* 38, no. 1 (1996): 47–81.

Davis, Jim, Thomas Hirschl, and Michael Stack, eds. *Cutting Edge: Technology, Information, Capitalism and Social Revolution.* London: Verso, 1997.

Dery, Mark, ed. *Flame Wars: The Discourse of Cyberculture.* Durham, NC: Duke University Press, 1994.

Ehrenreich, Barbara. *The Hearts of Men: American Dreams and the Flight from Commitment.* Garden City, NY: Anchor Books, 1983.

Ess, Charles, ed. *Philosophical Perspectives on Computer-Mediated Communication.* Albany, NY: State University of New York Press, 1996.

Furger, Roberta. *Does Jane Compute? Preserving Our Daughters' Place in the Cyber Revolution.* New York: Time Warner, 1998.

Green, Eileen et al., eds. *Gendered by Design?* London: Taylor and Francis, 1993.

Hale, Constance, ed. *Wired Style: Principles of English Usage in the Digital Age.* San Francisco: HardWired, 1996.

Harding, Sandra, ed. *The Racial Economy of Science: Toward a Democratic Future.* Bloomington: Indiana University Press, 1993.

Herring, Susan, Deborah A. Johnson, and Tamra DiBenedetto, "This Discussion Has Gone Too Far!: Male Resistance to Female Participation on the Internet." In Kira Hall and Mary Bulcholtz, eds., *Gender Articulated: Language and the Socially Constructed Self.* New York: Routledge, 1995.

Kroker, Arthur, and Michael A. Weinstein. *Data Trash: The Theory of the Virtual Class.* New York: St. Martin's Press, 1994.

Lupton, Ellen. *Mechanical Brides: Women and Machines from Home to Office.* New York: Cooper-Hewitt, 1993.

Menzies, Heather. *Whose Brave New World? The Information Highway and the New Economy.* Toronto: between the lines, 1996.

Mitter, Swasti, and Sheila Rowbotham, eds. *Women Encounter Technology: Changing Patterns of Employment in the Third World.* London: Routledge, 1995.

Mogensen, Vernon L. *Office Politics: Computers, Labor and the Fight for Safety and Health.* New Brunswick, NJ: Rutgers University Press, 1996.

Noble, David F. *The Religion of Technology: The Divinity of Man and the Spirit of Invention.* New York: Alfred A. Knopf, 1997.

——— *Progress Without People: New Technology, Unemployment and the*

Message of Resistance. Toronto: between the lines, 1995.

——— *A World Without Women: The Christian Clerical Culture of Western Science.* New York: Alfred A. Knopf, 1992.

Plant, Sadie. *Zeros + Ones: Digital Women + the New Technoculture.* New York: Doubleday, 1997.

Raymond, Eric S. *The New Hacker's Dictionary.* Second Edition. Cambridge, MA: MIT Press, 1994.

Rifkin, Jeremy. *The Biotech Century: Harnessing the Gene and Remaking the World.* New York: Jeremy P. Tarcher/Putnam, 1998.

Ronai, Carol Rambo, Barbara A. Zsembik, and Joe R. Feagin, eds. *Everyday Sexism in the Third Millennium.* New York: Routledge, 1997.

Roszak, Theodore. *The Cult of Information.* Berkeley: University of California Press, 1994.

Sardar, Ziauddin, and Jerome R. Ravetz, eds. *Cyberfutures: Culture and Politics of the Information Superhighway.* New York: New York University Press, 1996.

Segal, Lynne. *Slow Motion: Changing Masculinities, Changing Men.* Revised Edition. London: Virago Press, 1997.

Shiva, Vandana. *Biopiracy: The Plunder of Nature and Knowledge.* Toronto: between the lines, 1997.

Silverstone, Roger, and Eric Hirsch, eds. *Consuming Technologies: Media and Information in Domestic Spaces.* London: Routledge, 1992.

Slouka, Mark. *War of the Worlds: Cyberspace and the High Tech Assault on Reality.* New York: Basic, 1995.

Spender, Dale. *Nattering on the Net: Women, Power and Cyberspace.* Toronto: Garamond Press, 1995.

Springer, Claudia. *Electronic Eros: Bodies and Desire in the Postindustrial Age.* Austin, TX: Texas University Press, 1996.

Stabile, Carole. *Feminism and the Technological Fix.* Manchester: Manchester University Press, 1994.

Stanley, Autumn. *Mothers and Daughters of Invention: Notes for a Revised History of Technology.* New Brunswick, NJ: Rutgers University Press, 1995.

Stewart Millar, Melanie, ed. "WomanTech." Special Issue *WE International* 42/43 (Fall/Winter 1997).

Taylor, H. Jeannie, Cheris Kramarae, and Maureen Ebben, eds. *Women, Information Technology and Scholarship.* Urbana-Champaign, IL: Center for Advanced Study, University of Illinois, 1993.

Tomaskovic-Devey, Donald. *Gender and Race Inequality at Work: The Sources and Consequences of Job Segregation.* Ithaca, NY: ILR Press, 1993.

Turkle, Sherry. *Life on the Screen: Identity in the Age of the Internet.* New York: Simon and Schuster, 1995.

van Dijk, Teun, ed. *Discourse Studies: A Multidisciplinary Introduction.* Volumes 1 and 2. London: Sage Publications, 1997.

van Zoonen, Liesbet, "Feminist Theory and Information Technology." *Media Culture and Society* 14 (1992): 9-29.

Young, John E. *Global Network: Computers in a Sustainable Society.* Washington, DC: Worldwatch Institute, 1993.

Index

Illustration Credits

Figure 1 (page 21): Compaq Advertisement. Reprinted with permission of Compaq Canada Inc. (This advertisement appeared in *Wired*, June 1997.)

Figure 2 (page 99): "Nurses." © Steve Speer, 1998. Reprinted with permission of the artist.

Figure 3 (page 105): DNA Sluts. From All NEW GEN Interactive Computer Generated Image VNS MATRIX (Aust) © 1993. Reprinted with permission.

Figure 4 (page 129): Duke Nukem. Artwork © 1996–1998 3D Realms Entertainment. Reprinted with permission. (This advertisement appeared in *Wired*, September 1996.)

Figure 5 (page 168): Motorola Advertisement. Reprinted with permission of the Motorola Multimedia Group.

Figure 6 (page 170): *Women'space* cover, *Women'space* Fall 97/ Winter 98. Illustration by Juliet Breese. Reprinted with permission from *Women'space* magazine. Anyone wishing to reproduce Juliet Breese's illustration should contact *Women'space* at: (613) 256-5682 or diamond@womenspace.ca>